Central Asia and Kashmir
-Archaeobotany and Floristics-

Central Asia and Kashmir

-Archaeobotany and Floristics-

P.Kachroo

University of Kashmir,
Centre of Plant Taxonomy,
Srinagar, India

SCIENTIFIC®
PUBLISHERS

Published by
SCIENTIFIC PUBLISHERS (INDIA)

Jodhpur –

5 A, New Pali Road
P.O. Box 91
Jodhpur - 342 001 INDIA

Print 2026

ISBN: 978-81-7233-079-8

Visit the Scientific Publishers (India) website at
http://www.scientificpubonline.com

Printed in India

ACKNOWLEDGMENTS

Some of us in this university have been involved in studying cultural and botanical links with Central Asia. Recently we released a volume entitled "Central Asia and Western Himalaya- a forgotten link" (ed. G.M. Buth). Dr. Uppeander Dhar brought out recently an interesting study on phytogeographic analysis of flora of Kashmir Himalaya and showed how intimately it leaned on Central Asia. Conclusions from endemism by Dr. Virjee are similarly interesting.

The present contribution on "Central Asia and Kashmir-archaeobotany and floristics" is meant to be part two of Dr. Buth's book. It was largely compiled during my brief visit to the Central Asian Republics of the USSR. The first draft of this book formed the basis of my "Prof. P. Maheshwari Medal Award Lecture". Here I wish to express my gratitude to the then President and the Executive Council of the Indian Botanical Society for having selected me for this Award and thus honoured me.

I am grateful to the University Grants Commission and the USSR Academy of Sciences for financing my visit to the USSR. For library, herbarium and related facilities, such as local transport, provision of interpreters (for this also to the Academy), etc. I am thankful to the Directors of the various Botanical Institutes/Gardens at Dushanbe, Alma Ata, Leningrad and Moscow, particularly Prof. Isa Baitulin (and his Deputy Director), Drs. A. Takhtajan, A.A. Konnov, A.A. Ashurov, U.K. Mamatkulov and Igantov Michail and their colleagues at each place.

Thanks are due for encouragement to Hon'ble Shri T.N. Kaul and Dr. A.P. Kulshreshtha (Moscow), M.L. Kachroo, (New Delhi), Profs. P.N. Mehra (Chandigarh), and A. Takhtajan (Leningrad). Drs. A.R. Naqshi (Srinagar), B.V. Shetty (Kew) and U. Dhar (Almora) have helped me in a number of ways; and Shri Pawan Kumar, the Publisher, for his painstaking care. I am also thankful to Virjee, my student, for permitting me to use some of his unpublished data on endemism of the flowering plants of Kashmir.

To my daughter and son-in-law I owe special thanks for the gift of an electric type-writer which proved of great help.

To my wife, I dedicate this book in appreciation of her understanding and love.

New Delhi,
1 Jan, 1995

P. Kachroo

CONTENTS

CONTENTS

1

INTRODUCTION

Kashmir lies on the main route of migration between Gondwana and Angara continents and the flora of this region assumes special interest; and this lead Sahni (1938) to state that it shows close affinity with the Siberia flora than to that of the Peninsular Gondwana. He observed that *Glossopteris* flora in Kashmir reaches to within a few degrees of the Angara flora in the USSR; and that Kashmir was only an out-post of Angaraland (Sahni, 1926). He also (1935) had suggested that obvious relations with Angara flora especially as regards Kashmir where a transitional flora indicates northward migration cannot be over- ruled. There however was need for a critical comparision of the Siberian and Indian floras to assess the degree of affinity to clarify notions regarding the relations best.

Central Asia which in geologic times was an original divide between western and eastern groups of sea basin (the Atlantic and the Pacific) is of considerable interest from the view point of palaeoenvironment and human migration. It is also of interest in global palaeoenvironmental reconstructions.. (because) "important environmental changes took place in this region over the last glacial- interglacial cycle. It is possible that "climatic change in Central Asia was far greater than at other middle altitude sites". Howerer, "the magnitude of climatic change from interglacial to full glaciation conditions in Hindukush was broadly similar to that documented in major mountain chains in Europe, North America and South America. It may be recalled here that this region lies upwind from the high plateau of Central Asia (Porter, 1984).

In Eurasia the Angara flora is wedged in between the *Gigantopteris* and Arctocarboniferous provinces. In fact the Angara flora contains elements of all the three floras as well as that peculiar to itself. This synthetic nature is evident as Angaraland occupises a central position. The *Glossopteris* flora does contain stray elements of the northern type yet it is distinct and strongly characteristic of the southern type hemisphere.

A direct connection between Africa and Europe was in the Miocene, probably in the region of Balkans; and "this contact was accompanied by an intensive exchange of both plants and animals between Africa and Europe. Garson and Krs (1976) stated that the collision of the Arabian plate following its rifting from Africa in Eocene with Asia has resulted in the complex topography of Turkey and Central Asia (cf. Raven,1980). Prior to southward movement of Africa starting in Paleocene Africa and Asia may have been connected through Arabian Peninsula, Iran and Afghanistan might have laid between Magasay and Arabia (Fig.1) as a part of Gondwanaland (Raven 1.c.).

The flora of Middle Asia is regarded as a mainly autochtonous formation that had developed in the course of the reductional xerogenous transformation of the "Turkmenian" Palaeogene floras, as part of the floras of the islands and of the south coast of Tethys" (Kemelin, 1967). The strong sea floor spreading in the Late Phanerozonic time in the NW Indian Ocean caused a situation in which the whole Indo-Tibetan plate was close to the rest of Asian but "could not approach much more. The consequent overthrusting produced the Himalaya. "In Precambrian time Tibet and Gondwana India was one" (Crawford, 1974). Between the Gondwanic India (Peninsula of today) and western Australia lay the Tarim Basin Block which "appears to have moved sinistrically relative to Tibet... and its palaeo-position in Gondwanaland seems likely to have been adjacent to northern continental margin of Australia"(Fig.2).

The India region "appears to have separated from southwest Australia and adjacent Antartica at

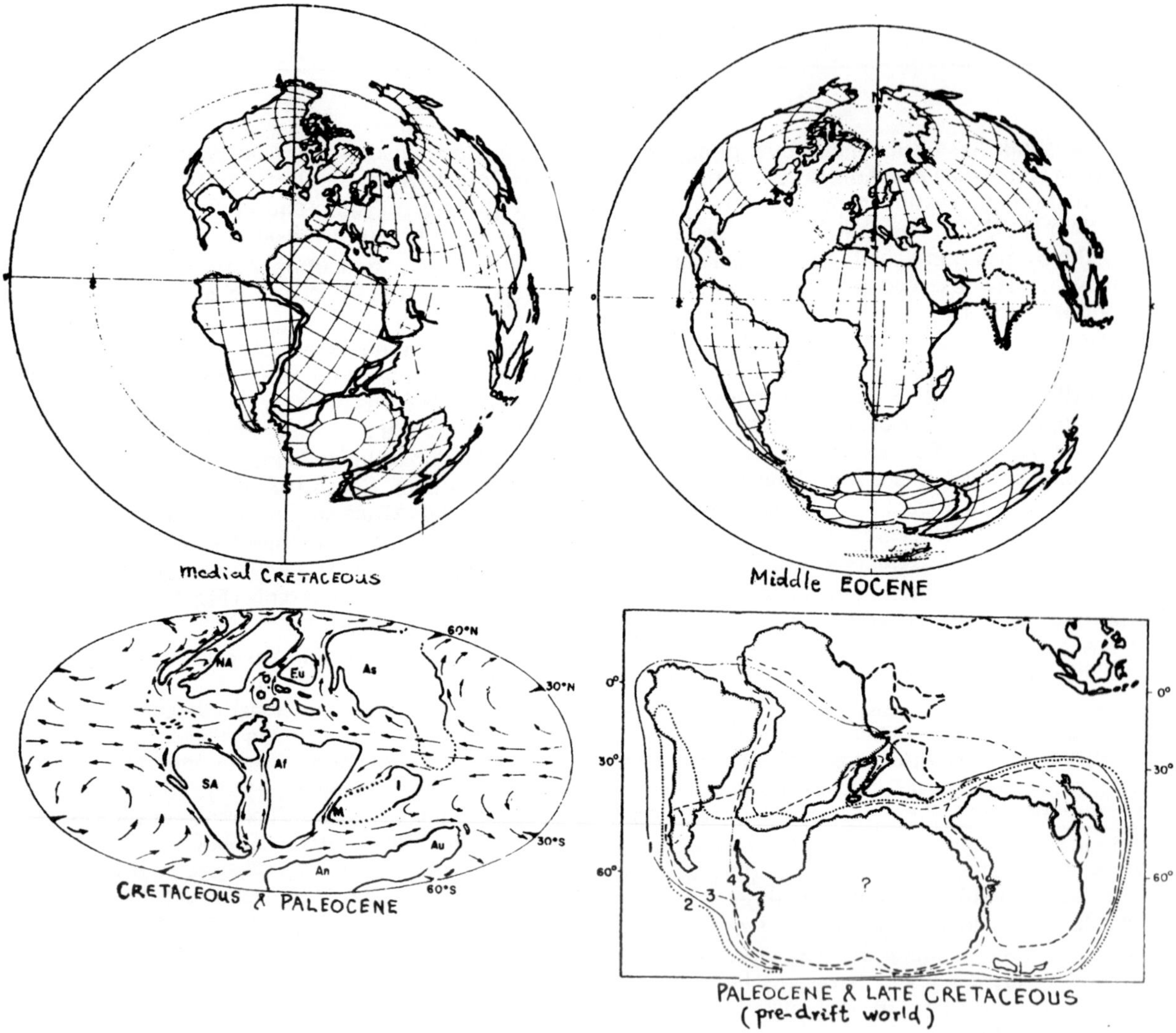

Fig. 1 : The world in medial Cretaceous, Cretaceous & Palaeocene, Middle Eocene, and Palaeocene & Late Cretaceous pre-drift world (cf Raven, 1980).

about 125 my B.P. (Veevers et al, 1975 , in Raven 1.c). Prior to 70 my BP Africa-Madgascar- Seychelles- India- Antartica-Australia were much closer than they are at present (Raven, 1980).

Dianelli (1922) suggested that Pleistocene uplift was of 2000 m in Kashmir. The Tibetan part of the Indo-Tibet may have started to "rise before the main Himalaya orogeny as the marine Eocene of Tibet passes up into Continental deposits (Crawfold, 1974).

The Indus suture line is regarded by Dewy and Bird (1971) as the relic of a closed ocean, "that which lay between Gondwanic India and Asia (Crawford, 1974). Crawford does not agree that this ocean was a wide one though India remained a part of Gondwanland until the Jurassic"... the oceanic crust -- existed only from the Permian to the Late Jurassic"... (Fig. 2) and that before the Permian no break down to the mantle existed" and the relative short period of crustal opening to the mantle was limited in regional extent." Thus India and Tibet are to be regarded for most of Phanerozoic time as one huge crustal unit, the southern half emerged, the northern submerged."

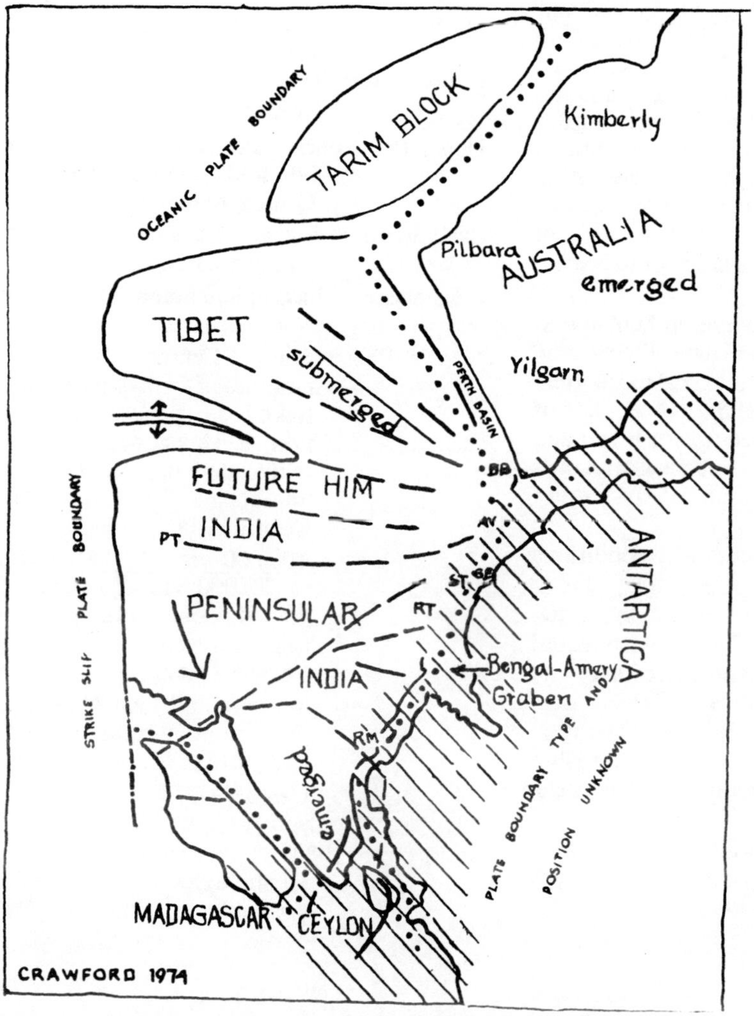

Fig. 2 : Part of Gondwanaland during Permian-Late Jurassic (cf Crawford, 1974).

2

ARCHAEO-PALAEOCLIMATIC PERSPECTIVE

Anthropology

Human occupation in Asia from 100,000 to 35,000 years ago embraces the geological climatological events of the third interglacial and initial ice advances of the last glacial period at a time when population movement was severely restricted by the eastern Europe, the Tibetan plateau and the diagonal mountain chains of Central Asia during movements of peak glaciation. These conditions blocked passage between Europe and eastern Asia save for narrow corridors not covered by ice. For this reason high altitude glaciated regions in western Asia and in the Himalya are poor in prehistoric sites (Kenneday, 1984).

Davis (1984) in his discourse on human distribution in the Late Middle and Upper Pleistocene has stated that Central Asia also experienced great cyclic climatic changes (events) as in other parts of the world and we have good records of lower and middle Palaeolithic occurrences but meagre records of upper Palaeolithic (Davis et al, 1980). Further that in:

Lower Palaeolithic : climatic conditions were interstadial or interglacial.

Middle Palaeolithic : not well dated; however Pakhamov et al (1982) in Tadjhikistan record that Middle Palaeolithic site experienced cool and wet "pluvial" conditions (in Davis, 1984).

There is reason to believe that casual correlations existed between climatic fluctuations distribution in human settlements mainly for Pleistocene and Holocene.

Kennedy (1984) in his interpretation on the hominid skeletal record (Fig.3) has suggested a few models which are enumerated below

Neanderthal hominid skeletal record

44,490 yr BP . . Shanider cave (Iraq), bones of child and adults.

35,000 yr BP . . Teshik-tas cave(Crimea)Amarkutan cave (Samarkand) earliest fossil Neanderthal in Central Asia.

30,000 yr BP . .Darri-i-Kur (N/E Afghanistan) also near *Homo sapiens sapiens*.

Upper Palaeolithic and Late Upper Palaeolithic hominids . . few and fragmentary in Central Asia.

Early terminal Pleistocene skeletons . . Serai Rai (Gangetic Plain).

Pleistocene . . incomplete human skull at banks of Narmada river.

Middle Pleistocene . . Central Aisa, lithic technologies of form of pebble tools and flake industries (was it Soan tradition ?). This appears to have been replaced by upper Palaeolithic tool tradition (45,000-35,000 yrs ago) in Central Asia.

100,000 yrs ago (last interglacial) to 70,000 yrs ago (onset of last glacial) stone working techniques, fully represented in West Asia and Central Asia.

In South and East Asia the pebble tool tradition continued well after the close of the Pleistocene (10,000 yrs ago).

Pleistocene tools in Kashmir (Sankalia, 1971)

Joshi et al (1984) . . intermontane penetration of technologies may have been posssible. Presumption .. a corridor between Central Asia, though Palaeolithic industries absent in Afghanistan!

Upper Palaeolithic culture of Central Asia had close ties with Middle Stone Age cultures of South Asia!

Sahni (1934) has remarked that the relics of Old Stone Age discovered at short intervals within the last few years focus attention upon northern India as an area of unusual promise for our knowledge of early man.

The discovery of Palaeolithic stone flake industries in three widely separated parts of northern

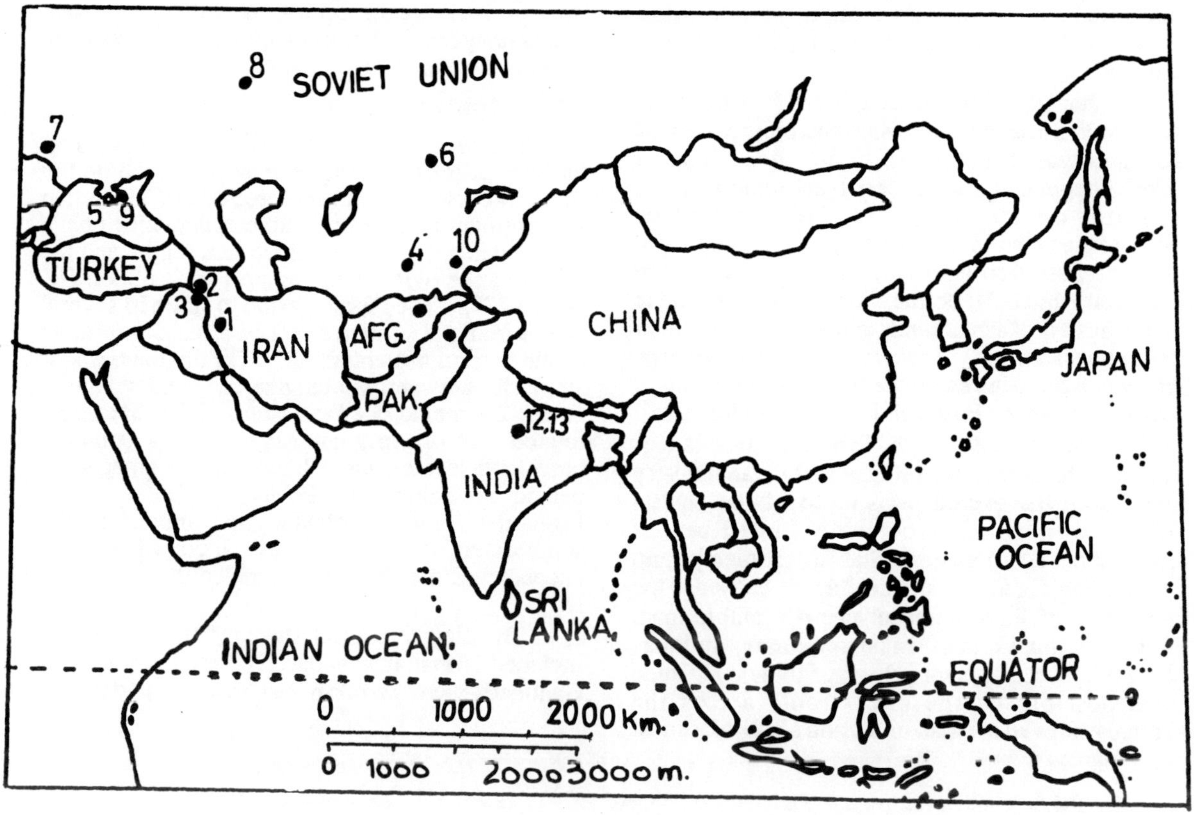

1 KERMAN SHAH
2 TAMTAMA
3 SHANIDAR
4 AMAN-KUTAN
5 KHK-KOBA
6 MALTA
7 MOLODOVA
8 OUNDORY
9 STAROSELYE
10 TESHIK-TASH
11 DARRA-I-KUR
12 SARAI NAHAR RAI
13 MAHADAHA
14 SOAN VALLEY

Fig. 3 : Major sites of Pleistocene hominid remains in Central and South Asia (cf Kennedy, 1984).

India : southwest of Rawalpindi (Chitta, Pakistan), few kilometers east of Srinagar (Pampur[1], Kashmir) and at Kargil[2] (Ladakh) are revealing with respect to early human activities (Hawkes et al, 1934). Kargil lies beyond the main Himalayan range on the ancient trade route over the Zojila connecting India with Central Asia, Tibet and China (Sahni, 1934). Stone implements of Lower to Middle Palaeolithic and Middle and Upper Palaeolithic cultures have also been discovered near Pindigeb (Attock, Pakistan).

Sahni (1934) further stated that "round about Middle Pleistocene time when the main valley of Kashmir was still occupied by the great 'Karewa lake" interglacial man of about the same stage of cultural development - as Neanderthal or Mousteriian man in Europe and as Peking man is in the Far East flourished (a) in the plains of the northern Punjab (b) on the shores of Karewa Lake in the heart of Kashmir and (c) just across the Great Himalayan range.. *"indicating contact between early human cultures on the two sides of the main Himalaya and the Pir Panjal range"* (italics mine).

In this connection de Terra's remark is significant, that traces of a prehistoric human industry have been discovered even north of the central Himalayan range on the borders of Little Tibet ! In fact Sahni (1934) asserted that the Himalaya and the Pir Panjal range could not "act as barrier to the migration of Palaeolithic or even Neolithic man. Thus long before man conquered the ocean intercourse between ancient culture of India and China was possible by the direct route across the Himalaya" even in Palaeolithic and Neolithic times!

de Terra & Paterson (1939) believed that the early man came from Peninsular India to Kashmir since Palaeolithic antiquities are lacking in the valley and that early man did not discover Kashmir until the end of Ice Age. But this is not tenable as Sahni (1934) has stated that "it might be difficut to explain crossing of Zojila (3,800 m) by the Palaeolithic man as he was primarily a hunter, yet "migration across these passes over them were then presumably not so high as to offer a serious obstacle to primitive man".

Archaeology

The new way of life began in the uplands on the foot hills of the Zagros mountains and on the Anatolian plateau, a little before 5000 BC. In Indian subcontinent traces of neolithic communities have survived mostly in the northwestern region and the Deccan.. the oldest remains go back to 3500 BC (Jha, 1977). "The Aryans came to India in several waves from the western direction around the middle of the second millenium BC". On the the basis of similarity between the languages, Sanskrit, Latin, Greek, Germanic, Slav and Roman it has been postulated that the original Aryans had a common homeland located most likely in the great steppeland stretching from Poland to Central Asia". From the Iranean tableland they moved in the southeastern direction towards India where they encountered the city civilization of the Indus valley".

The first territorial states of the Aryans[3,4] included 16 mahajanapadas. Of these Kamboja and Gandhara were in northwest in present day Af-

(1) The flakes found in the lower Jhelum terrace (Pampur) may represent a late Palaeolithic or Protoneolithic culture (de Terra & Paterson, 1939). At Burzahom the flakes associated with pottery bearing layers are products of an advanced Stone Age technique, and such Neolithic sites were contemporaneous with the soil formation and "the neolithic soil is a loess of post glacial age".

(2) The Kargil man may be attributed to the great family of Palaeolithic flake industries era known also in Chitta and Pampur.

(3) The Rig Veda mentions five seasons which shows that Aryans possessed some knowledege of seasons which promoted agriculture.
Kashmiris even now recognise six seasons; "From the very little we know .. they (the Aryans) were not of such an advanced culture as people whom now we term the Harappan people" (Mackay, 1976).

(4) Phallic worship ... seems to have obtained among the Harappans.The three Harappan cults, tbose of Mother Goddess, proto-Shiva and the linga later became features of Indian religion. Kashmir still strongly clings to the tradition. Shiva is 'one of the oldest Indian Gods and his worship is dated from the pre-historic period" (Mackay, 1976) and this was not the name given to Him by the Harappan people. In fact today Shiva is known "to have a thousand and eight names". But on two seals, a spring of flowers or leaves rising from the head between the horns (of Shiva) strongly suggest that the figure so ornamented was a fertility or vegetation god, analogous with Shiva, who personifies the reproductive powers of Nature".

ghanistan and Kabul region respectively. Taxila was a leading city.

"According to tradition Kashmir was a part of the Maurya empire and Ashoka built the city of Shrinagara". Khotan in Central Asia is also said to have come under Mauryan sway; the Tibetan sources even refer to Ashoka's visit to the region.

The Late Cenozoic sediments are reported to be well preserved in Kashmir valley and that during this epoch the Pir Panjal range rose by a few meters giving rise to a vast lake by blocking natural drainage. Further uplift of this range resulted in erosion process which exposed large section of lake sediments resulting in three lithologic units (Kusumgar et al,1984):the lower and upper Karewas and the loess.

These Karewas contain a wealth of information on prehistoric climate ranging from Miocene to Pleistocene (Bhatt, 1976) and early man in Kashmir. In this connection reference is made to Neolithic culture of Burzhom. The earliest evidence of agriculture so far discovered in India may be stated to be 2300 BC as is apparent from radio-carbon dates of the Neolithic sites in Kashmir valley, (Fig.4) Mysore and Andhra Pradesh . . . the Neolithic period in India commenced about 2500 BC and continued to 1000 BC (Agrawal & Kusumgar 1984, Agrawal,Gupta & Kusumgar, 1969) . . . between 2300 and 1700 BC Harappan culture flourished in Sind, Punjab,Rajasthan and Saurashtra ...emergence of Iron Age in India is ca 1000 BC.. and ended about from 450 BC. The advent of Aryans into India is now believed to date from the late Chalcolithic and early Iron Age" (Vishnu-Mittre,1970).Agrawal & Kusumgar (1966) had stated that the Neolithic in Kashmir valley extended to 1500 BC and according to Vishnu-Mittre (l.c.) the early evidence of farming are "Undoubtedly of Neolithic age (Fig.4) dated in the valley as 2735 ± 120 to 1725 ± 95 BC. The extra territorial affinities of the Kashmir neolithic culture have been lucidly discussed by Dikshit (1982). The affinities lie with those of east Asia (part of north China,east Siberia,Manchuria,Korea and Japan). The novel semi-subterranean dwelllings characteristic feature of neolithic culture in Kashmir and China are shared by Swat valley (Pakistan). Dikshit (1982) gives the following characteristics of neolithic:

1. Semi-subterranean dwelling pits, storage, pits, fireplaces;
2. variety of bone tools including harpoons;
3. neoliithic stone implements (Fig.5) includ ing crescentic and rectangular knives (harvester) with perforation[5];
4. ritual burials of animals, especielly dog; and
5. hand made grey pottery. "the shape and form of handmade Burzhom" pottery have no parallel in North China"

The characteristics of other related areas are as follows:

Neolithic of China : was characterised by microliths in Gobi tradition and brown pottery with comb making and bone artifacts.

North China: (1) sub-terranean dwellings; (2) storage pits; (3) notched and crescentic rectangular stone knives with perforation; and (4) bone points including harpoon and hand made pottery.

Eastern Siberia: (1) the neolithic people introduced pottery crude in begining , later with textured design; (2) tools more simple and included flake scissors,half ground axes and adzes,there was elaborate fishing technology;(3) pit dwelling had roofs raised on pillars; (4) cemeteries ware present; (5) the people lived on fishing and hunting (Treistman 1972); and (6) human skeletons were stained with red ochre.

Manchuria (Weston 1971) : (1) neolithic pottery ; inpressed and black painted ; (2) vessels in grey ware ; (3) coarse ware were perhaps linked to sub-neolithic Gobi tradition; and (4) stone axes

5. "The stone knives-harvesters with peforation found in North and Central China with Yang Shao and Ling Shan complex were also present in Jomon neolithic phase in Japan and Korea. The distribution of this particular type along with other traits anticipates cultural connection between Kashmir, China and further east" (Dixit, 1982)

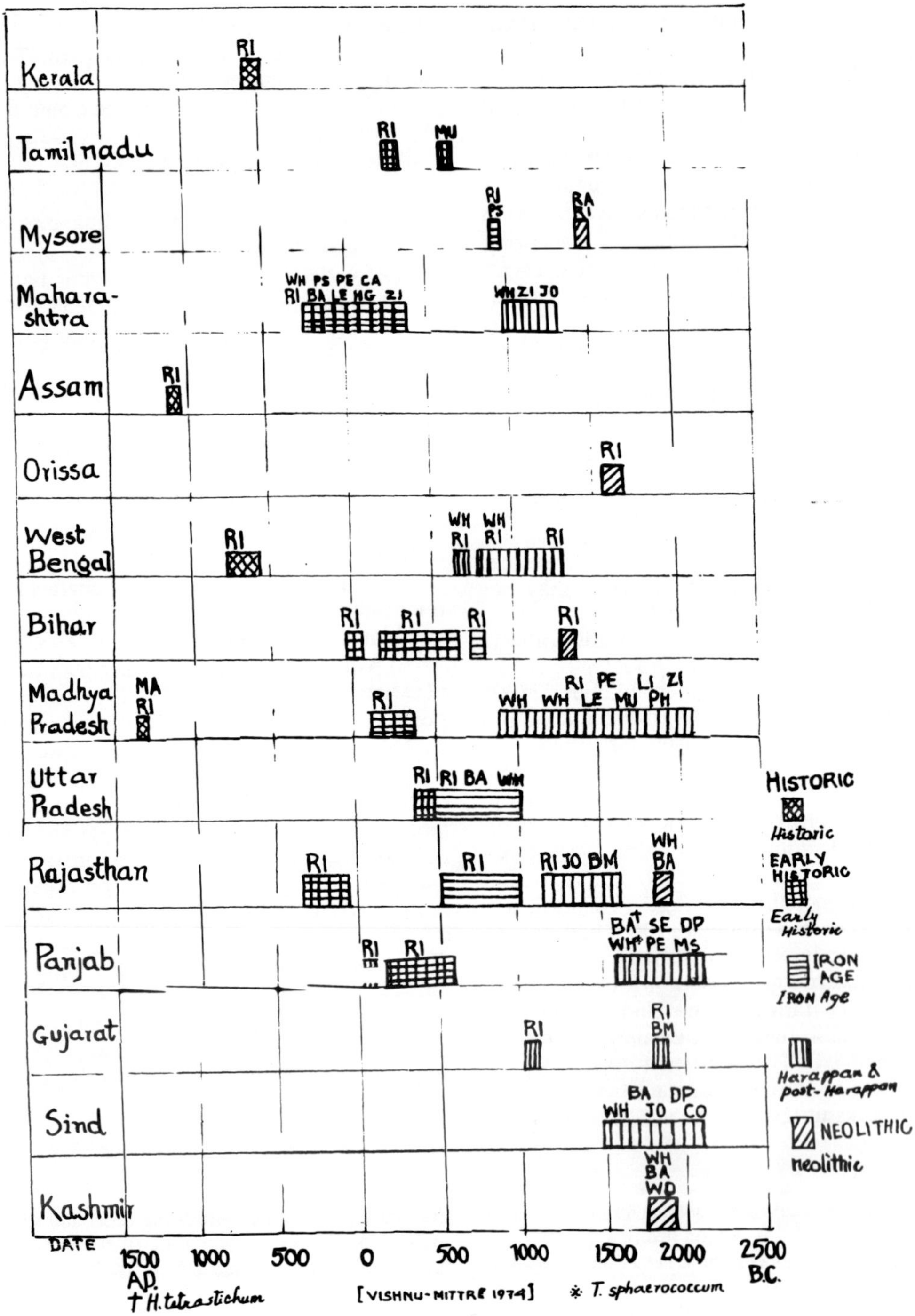

Fig. 4 : Distribution of cultfivared plants in space and time in the Indian subcontinent (cf. Vishnu-Mittre, 1974).

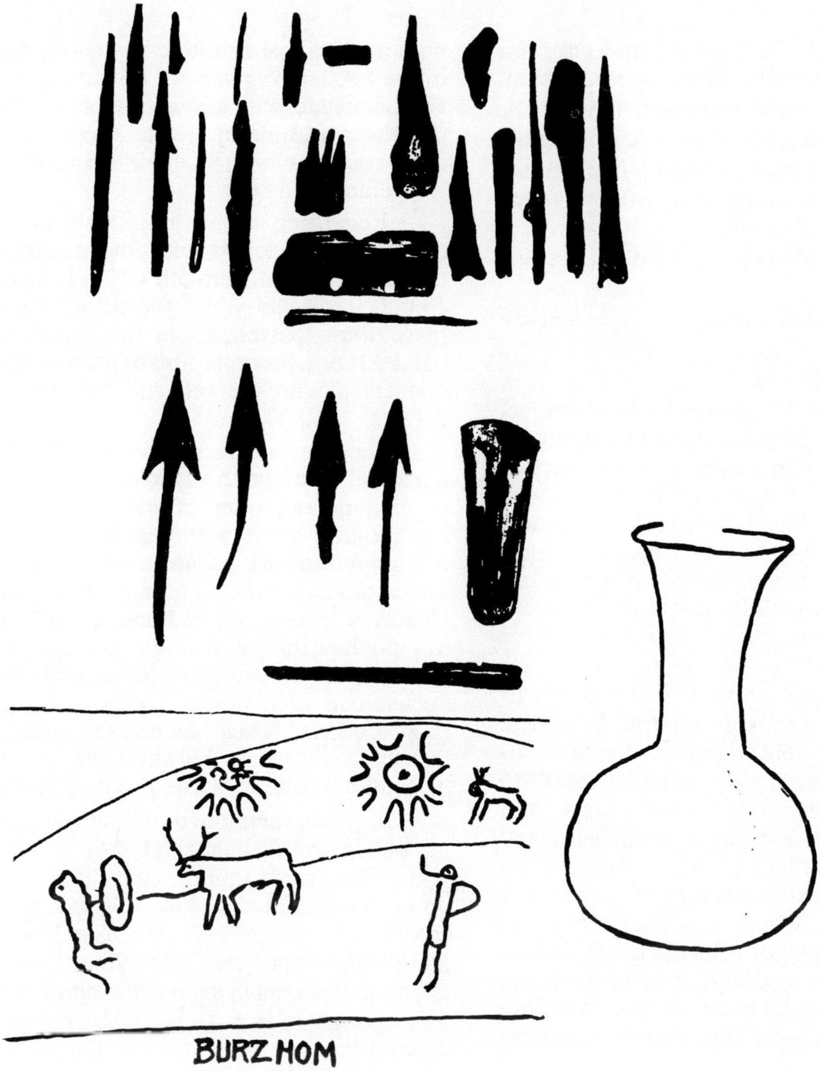

Fig. 5 : The tools excavated from Burzahom

Legend to Fig. 4

BA barley *Hordeum*
BM bajra *Pennisetum*
CA caster *Ricinus*
CO cotton *Gossypium*
HG horse gram *Dolicos*
JO jowar *Sorghum*
LE *Lens*
LI linseed *Linum*
MA maize *Zea*
MS melon *Cucumis*
MU mung *Phaseolus*
PE pea *Pisum*
PS *Paspalum scrobiculatum*
RI rice *Oryza*
WH wheat *Triticum*
WD Weeds
ZI *Zizyphus*
DP date palm *Phoenix*
PH *Phyllanthus*
RA *Eleusine ragi*
SE Sesame *Sesamum*

sickles, flake tools etc. A kind of primitive stone ploughshare and querns with rubbers were also seen.

Soviet Central Asia[6] (Mongait, 1959) :"The neolithoiic complex is more of west Asian origin".

(1) Amu Darya (between 4 and 3rd millenium BC surface dwellings with wooden frame and central fire places ;

(2) flint knives scrapers , polished axes and arrows;

(3) pottery flat vases bearing thread partern and painted , and

(4) hunting and fishing .

Dixit (1982) sees "an apparent idea of a movement of integrated neolithic culture[7] from the nuclear area of China to Kashmir ... the entry to Kashmir was through Gilgit and Serhad and then along the foot hills of Kun Lun ranges. Other route was along the river Khotan up to center of Tarim". About Swat Dikshit remarked that it was a late extension of Kashmir neolithic except for its dwellings[8] . Why could not the migration follow the reverse route?

The circumstantial evidence that presents itself is that from a 'nucleus' in Central Asia there could have been a two-fold migration , one to southwest (thence to Kashmir) and one east- north to China- Manchuria-Siberia. In Central Asia a from of neolithic culture is revealed at Amu Darya and west Kazakhstan (Mongait,1959). In fact Dikshit (l.c.) has also stated that "the entry to Kashmir was through Gilgit and Sarhad and then along the foothills of Kun Lun ranges" .But did neolithic man trod these routes into Kashmir, I have no 100% proof yet faith that he did come thence. There are historical facts and cultural links. I have no quarrel with the idea that neolithic of Central Asia may be of west Asian origon but I do not agree that there was no contact with eastern migrants (Dikshit calls them North Chinese) or that North China affinity alone counts as far as Kashmir is concerned we have little clues even here .

Regarding archaeological finds in Tadjik depression "25 levels with lower palaeolithic artifacts were fixed from 5th - 12th palaeosols from 13,000 - 850,000 yrs ... the palaeomagnetic data place these palaeosols in the brackets of 750-850,000 yrs ... the main sites of palaeolithic sites are connected with the 6th and 5th palaeosols (ca 200,000 - 130,000 yrs)".

The main artifacts occur directly embeded in palaeosols, the stone tools at bottom and middle part of climatic optimum of palaeosols. But no archaeological layers with hearths bones, structures for living etc" are found in any site. "However, archaeological finds in other sites, 54 km off Dushanbe, reveal broken bones of small ungulates but no hearths or storage pits.choppers. side scrapers, hollow and denticulate tools, flakes and "clactonian type *quartair d'orange*, the wedge shaped debris". Thus one can say with certaintity that the "industry of this site represents the lower palaeolithic culture.. a true pebble culture".

Rig Veda mentioned cultivation of yava (1,53) i.e. barley and of dhanah (1, 16) i.e. any kind of grain. This word unfortunately is confused with dhanya which means rice. Considering earliest colonisation of RigVedic Aryans in NW India, this confusion should not arise since cultivation of paddy in this area is against the agrobotanical history of origin of rice. Thus the absence of the name of vrihi (the boro rice of lower Bengal) in the Rig

6. The pre-historic people of Central Asia were in final stages of the Stone Age when the 6th millenium BC had just dawned (Ranov, 1982). South Turkemenia saw at the time the beginning of agricultural societies of the Djeitun Culture and animal husbandry was wide spread in the area. This was also an important era in the Hissar neolithic culture which spread in the foothills of Kopet-Dagh and in the desert parts of Soviet Central Asia (Masson & Sarianidi 1972; Shnilerjan 1980 cf, Ravov 1982) Considerable "similarity occurs in distribution of this 'pebble technique' (and economic basis also) between the Hissar neolithic culture and the final Soanian and post-Soanian in North Pakistan and India (Gupta 1978, Tanov 1968), and the tools collected ... at Hazar-Sum in Afghanistan may belong to the Hissar culture complx".

7. This culture apparently had little or no contact with western chalcolithic cultures (Basu and Pal, 1980). Alchin and Alchin (1968) have also observed that .. though this culture must have coexisted with the pre-Harappan and Harappan developments in the Indus valley and Panjab, there is no indication that the two enjoyed any contact what-so-ever. The question of relation with the other Indian Neolithic groups is more problematic.

8. Buth and Kaw (1984) stated that "there was continual if not regular exchange of man, materials and ideas between Kashmir and NW India from the days of yore; or there was regular ex-change as the food material was not indigenous, it is of West Asia origin".

Veda (cf Das, 1944) is not surprising. In fact rice may have been 'cultivated' in Bengal by the aborgines before Aryans invaded the land.We have archaeological evidence of "colonisation of Aryans into Bengal (Rajar Dhipi) upto 1100 BC only" (Fig.6).

Kashmir has assumed importance with respct to early human settlements in India with evidence that early man (Stone Age) lived in Pahlgam on the banks of Lidder (Valley) river(alt 3500 m) during the first interglacial period or during the second glacial dated by Sankalia (1977) to ca 500,000 years ago. "After this period the Himalaya began to rise". At the same time, period Early man existed at five sites in the Punjab hills (Sankalia, 1.c).

"Except writing of which we have no evidence so far, the early Kashmiri inhabitants of Neolithic culture of Kashmir valley had everything essential for living comfortable life" (ca 4000 yrs ago). These early Kashmiris wore clothes of skins and probably woven and pressed wool as indicated by the occurrence of large and small points and needles and scrappers of various kinds among stone tools" also these skin fur and woollen clothes might be like those worn by the Central Asian People". Sankalia established contact with Iran as well as Central Asia through sporadic but 'maintained throughout historic times".[9]

In a recnt study Yong (1984) stated that silk industry has a history of more than 5000 years. During 770-476 BC the silk may have reached Eurasian hinterland through nomadic tribes and thence spread to Europe via Scythians; no doubt there was not a regular trade route. To India the silk must have reached by 4th century BC since silk, cinapatta (chinese silk) is mentioned in Arth-sastra and 'ancient Indians already knew that silk was a secretion from a kind of insect'. The Silk Road as it is known was first opend in "middle of second century BC".

Let us consider the following dates:

Pre-Indus or pre-Harappan . . . . ca 3000-2000 BC
Indus or Harappan . . . . ca 2500-1500 BC
Burzahom (neolithic) . . . . 2500-1800 BC
Descent of Aryans into Punjab which Aryans found already in occupation of allïen people (Das, 1944) . . . . 3000 BC (Bhargawa, 1971).

It is not known for certain if the Rgvedic Aryans who enterd Kabul colonised Kashmir since in the list of rivers "in the Nadi stuti Hymn (Rg Veda 10,75)" and elsewhere we find (only) the names of the Ganga, the Yamuna the Sarayu and Sarawati .. of the western tributories of the Indus we find names of Kabha (Kabul river), the Surastu (Swat r), the Krumbu (Kurram r) and the Gomati (Goma r) (Das, 1944). There is no mention of any of the rivers between Kabul and Srinagar. But one might presume that like Pre-Aryan land of five rivers, Kashmir had an occupation. But what stage of civilization -we have no clues.

Regarding the earliest civilisations enumerated above we are still not sure what to call them. Thus Sankalia (1977) poses the question; did Pre-Aryans or Aryans represent Indus civilisation; were non - Aryans Harappan or Pre - Harappan; did vedic period represent Indus civilization? The main difficulty is that due to extreme inferiority complex our scholars are obsessed with comparative European dates. Notwithstanding that the Godavan flake furnishes evidence of the existence of man at a much earlier period than Europe (Osborn, 1911).

Palaeoclimate

Dadonov & Lomov (1984) have stated that "the age of most ancient elements of loess soil strata indicate that loess accumulation in Middle Asia started at least 2 m y ago, registering rhythmic climatic oscillation at this time". According to Dodonov (1984) the high terrace (600) 800 m above the level of the Central Asian foot hills "probably correspond to the Late Pliocene; alluvial terraces with 300- 400 m of elevation correspond with Eopleistocene deposits". The most ancient (3.5 m y) boundary in southern Tajikistan is Upper Pliocene.

The paleosols of Late Pleistocene and Middle Pleistocene are dated 130-200,000 yre (Dodonov & Penkov, 1977). The buried paleosol has compli-

9. According to Jha (1977) during ca 558-529 BC there were two routes from Persia to India (1) through the deserts of Gedrosia later followed by Alexander on his return from India, (2) Persia to Bactria, Sogdia and Sythia via Kabul valley, a region inhabited by Indian tribes and included in India, thence to Indus valley this route was followed by Alexander to invade India (see also Fig. 7).

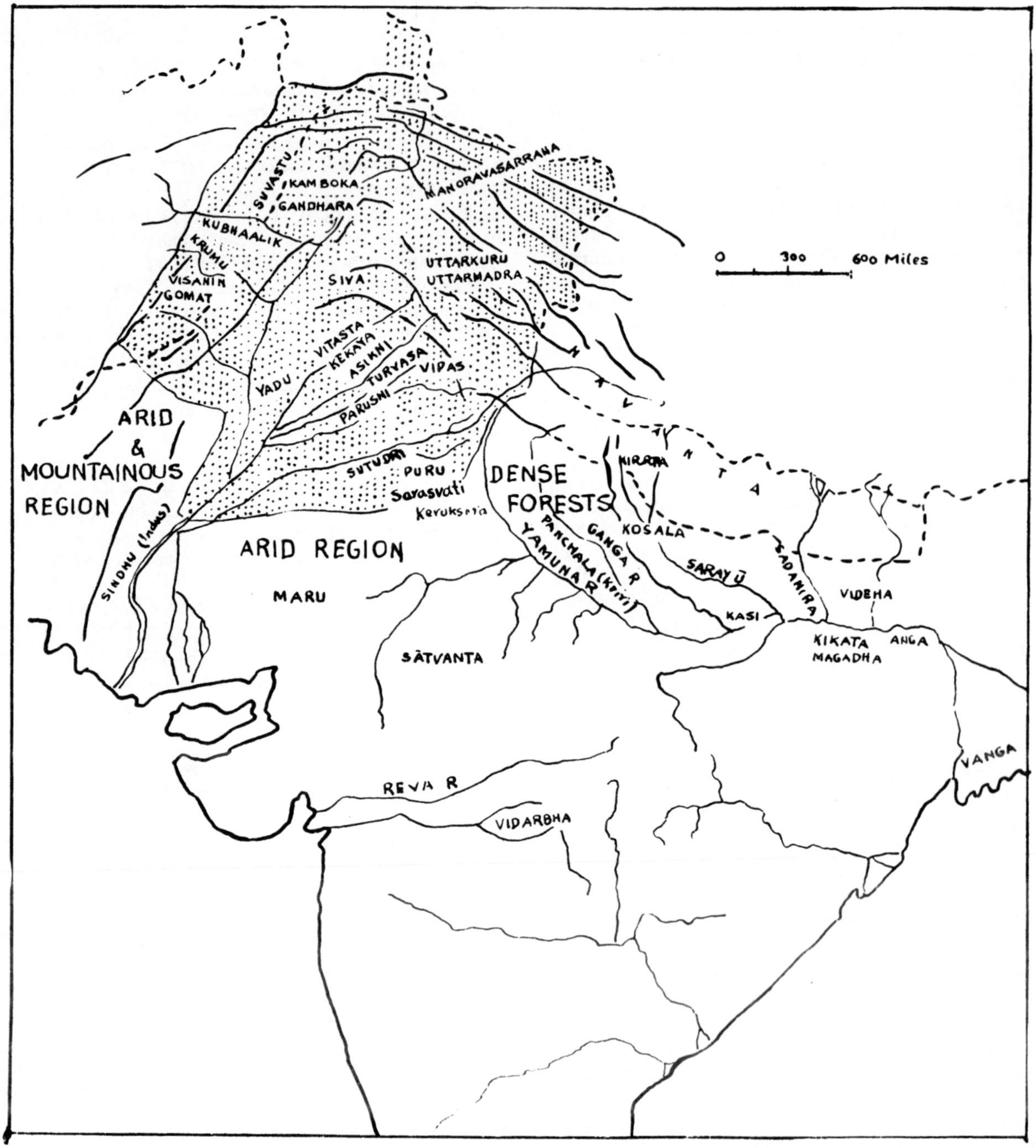

Fig. 6 : The first home of the Aryans in India (shaded areas) (cf Jha, 1977).

cated representing a pedocomplex composed of soils of different properties since these were formed in different climatic conditions; and that the ecological scheme might have been, "rise in temperature - - climatic opimum - fall in temperature" (Lomanov & Ranov, 1984).

Mitra (1927) has stated the region occupied by the Himalaya was once occupied by the Tethys Sea (Mediterranean Ocean) separating India from the Eurasian land-mass in the north. Into this Tethys India which lay south of equator sent two promonotories: the Kashmir promonotory on the NW and Assam promonotory on the NE. Wadia (cf Sahni, 1934) believed that "*Kashmir was for sometime connected to Eurasia by isthmus* (italics mine). From the Permian to the end of Eocene the bed of the Tethys was gradually being filled up with accumulating sediments and with the dawn of Tertiary era the ocean floor was loaded . . and were born the mountain chains bestween the northern and southern land blocks" . With the final phase of uplift of the Himalayan range the karewas of Kashmir assumed their present state (Fig.8).

Two outstanding characterstics are evident: the zone arrangement of rock formations from SW to NE with progressive complicated structures and

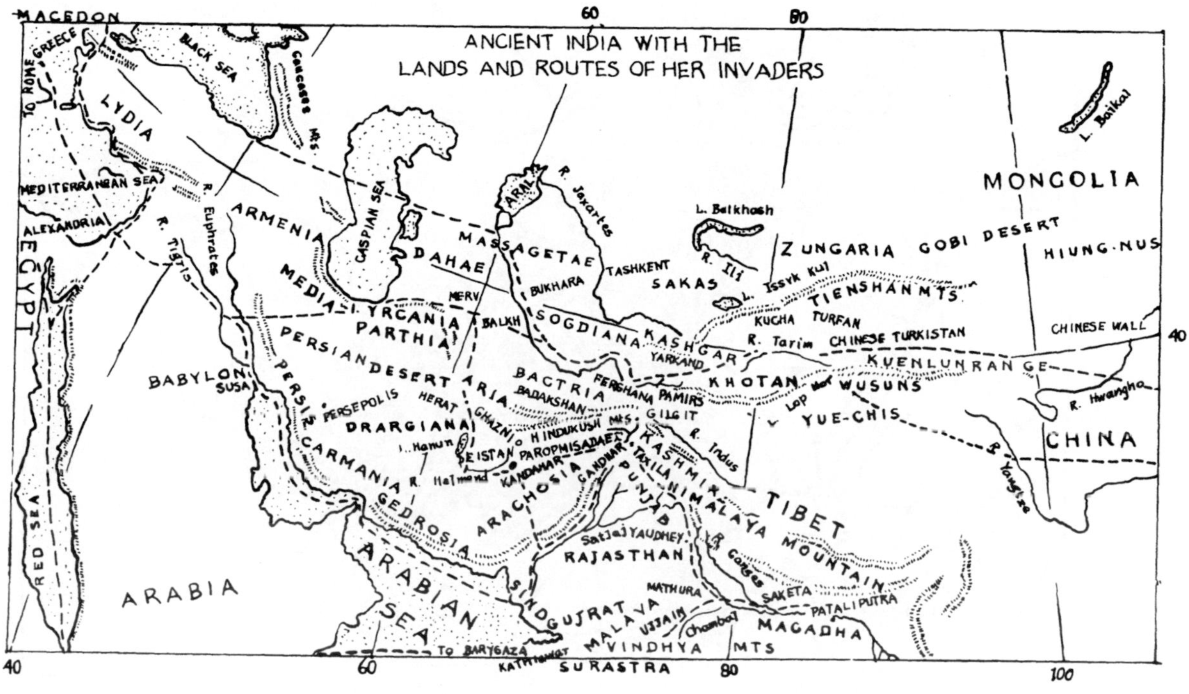

Fig. 7 : Ancient India with lands and routes of her invaders (cf Ojha, 1968).

conversely the progressive uplift and folding towards the Indian plains; this process has profound changes in palaeogeography of Kashmir and this is also reflected in the drainage pattern in the area.

An interesting event of geology of Kashmir is the Ice Age. The Sind and Lidder valleys experienced quaternary glaciation which de Terra and Paterson (1939) divided into four glacial and three interglacial epochs; of these the first two glaciations were more intense. The interglacial pulsations on ice front were more cvident in late stages than in the early because of erosion and weathering (Fig.9).

The second glacial period saw advances, the third period four advances and retreat halt and four advances in the fourth period with several retreat stages. The second interglacial period experienced oscillation of climate. Compared with interglacial periods,glacial periods were much shorter and interglacial erosion was much greater than during the glacial period. The Kashmir basin experienced glaciation in Pleistocene and this is summarised below (cf de Terra & Paterson. 1939).

Lower Pleistocene.. first glacial and interglacial stage. separated by angular unconformity or an erosion break from MP.

Middle Pleistocene (MP).. second major

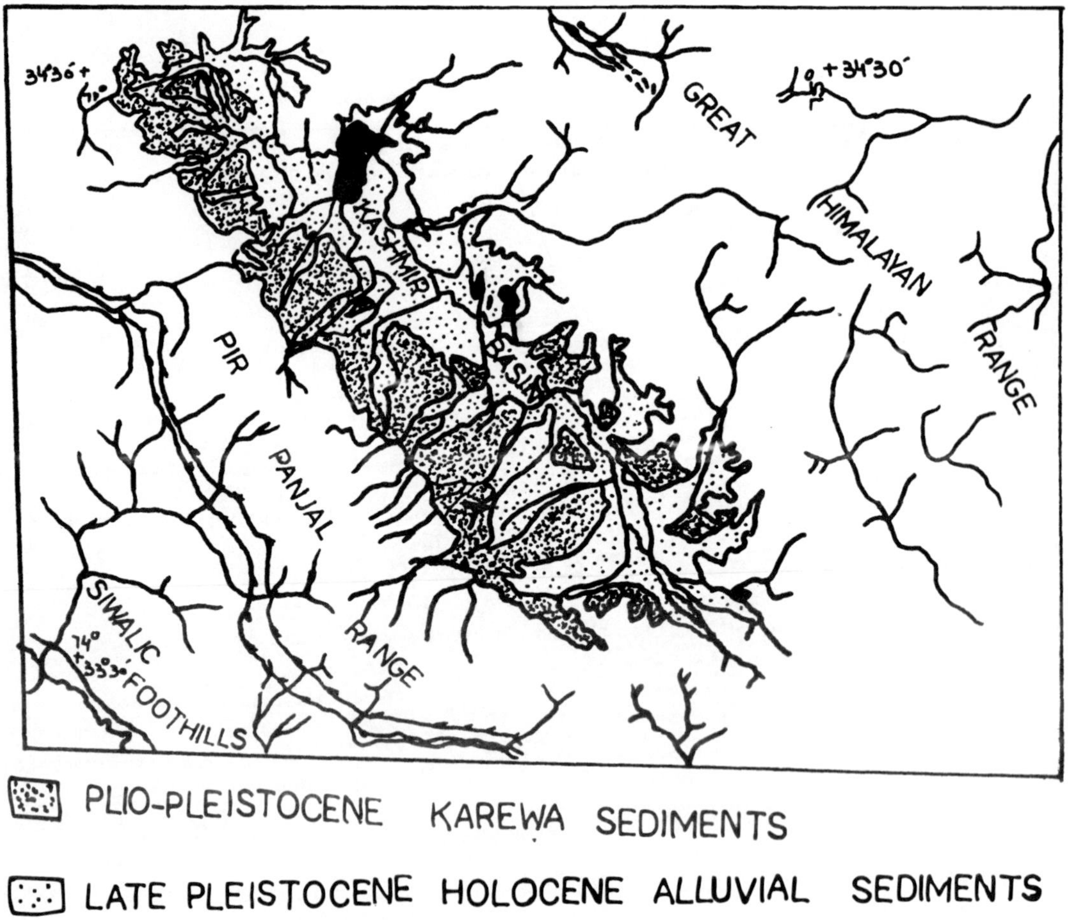

Fig. 8 : Kashmir basin showing distribution of the kerewa sediments.

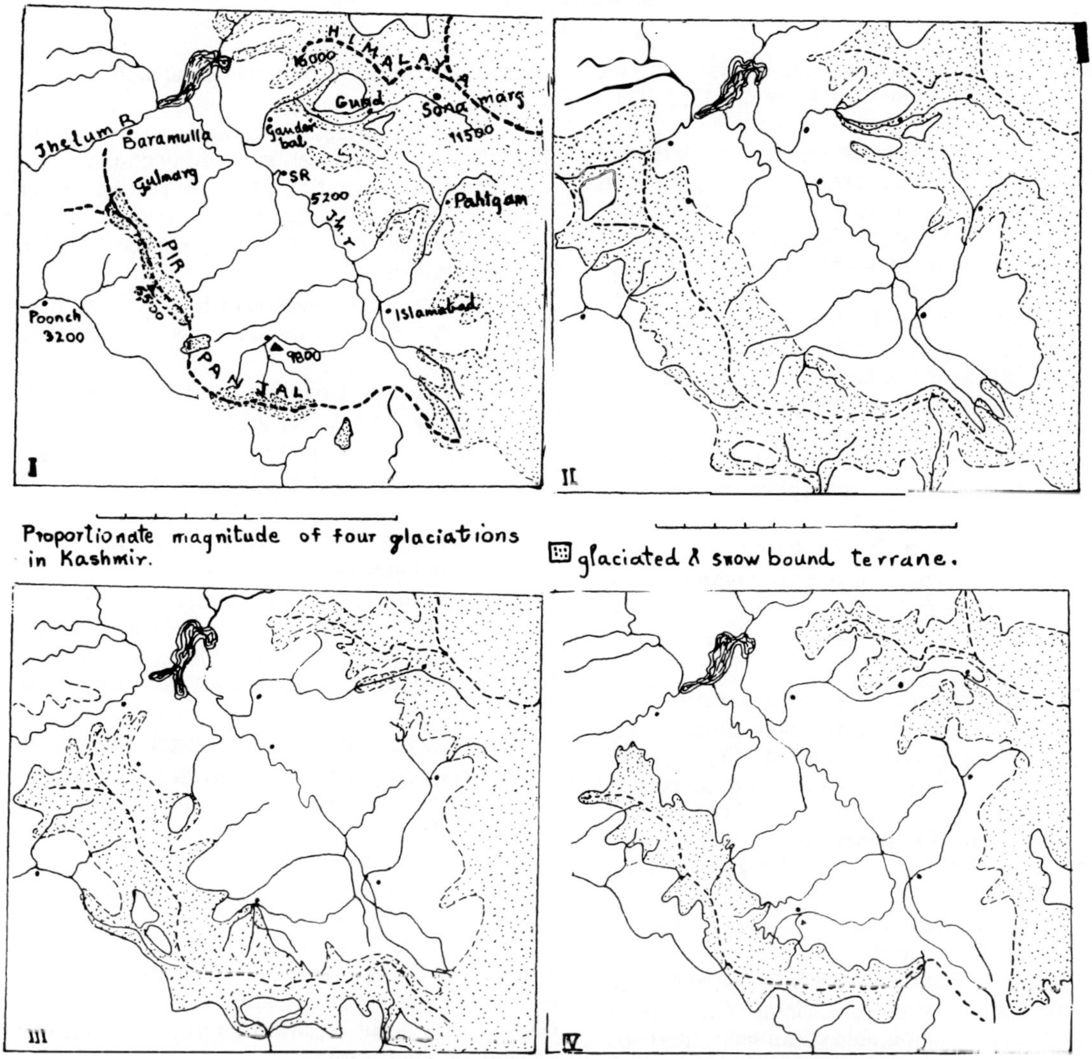

Fig. 9 : Proportionate magnitude of four glaciations in Kashmir (cf de Terra & Paterson ,1939).

glaciation, followed by long interglacial stage; separated from UP by unconformity through formation of boulder fans and thick glaciofluvial deposits followed by subsequent erosion.

Upper Pleistocene(UP) . . last two glaciations and an inter glacial period, the fourth glaciation least significat; separated from Holocene by final retreat of mountain glaciers and lifting of snow line to its present position.

Holocene or post-glacial . . continued ulift of entire Himalauan tract.

The dependence of Pleistocene glaiation on the diastrophic character of a mobile mountain belt is also known in Central Asia and the Alps.

In contrast to Pir Panjal the flora of Nanga Parbhat depicts a much more arid type of climate reminiscent of Central Asia (Troll,1938).

Major changes in flora took place in glacial periods. Once the aridity began to expand on a continental scale, it offered a new habitat and stimulus to evolution both to the derivatives of the tropical floras that flanked to the south,but also of the temeperate ones that bordered it on the north. Species were recruited and peobably still are being recruited from both of these areas and being recombined to make up the modern plant communities we see at the present time" (cf Raven, 1971). Thus during Pleistocene the Central Asian mountains and the Himalaya were "linked with the more extended Arctic Girdle". It was easy for the species to migrate.In fact the species complexes exisiting in Tertiary could be ancestors of all the xerophytic taxa and schizo-endemics; possibly alpine flora could seek their ancestors in the same ensemble. However, the alpine and nival flora growing today in more restricted alpine stands has partly or totally lost the ability to,or has had too little time to difereniate. "The exchange of flora between mountain ranges which occurred during the glacial period in fluctuations is one of the main reasons for the great similarity of the alpine and nival floras of the Asian mountains" (Breckle,1974).

Pleniglacial period in The Zaribar

Freitag (1977) has discussed the period based on pollen record as follows:

(1) periods with predominant semi- desert vegetation:

The surroundings of the lake Zaribar were dominated by Artemisia semi desert, probably associated with woody chenopods and a high proportion of *Ferula,* probably concentrated on eroded rocky slopes. Pistacia atlantica woodland of very open type occurred in ravines and other favourable habitats". In the outer belt of lake occurred halophytic vegetation mainly of chenopods. The lake itself was of fresh water. *Tamarix, Salix* and *Hippophae* formed shrubberies and low gallery forests in and low gallery forests in and along the river beds and water courses. Small Acer stands abounded in more favourable conditions, "perhaps in higher parts of the catchment area".

Quercus may have been present scattered in most humid habitats but considering the presence of *Artemisia,* chenopods and *Ferula* ; the *Quercus* pollen stems from medium distance transport and produced by Quercus woodlands on the other side of Zagros mts.

Climatic deterioration may have led to increase in chenopods in the semi-desert communities and to the complete disappearance of trees. Yet thermophilous taxa as *Tamarix, Salix, Lemna, Butomus, Bongardia* etc., did occur indicationg increasing aridity rather than a decline in temperature as decisive factor for the disppearance of most trees. "*Pistacia* and *Acer* may have survived the hardest period of Pleniglacial period in the area itself or very close to it".

The Pleniglacial vegetation around lake Zerober finds its closest present day counter -parts in the vegetation of the higher parts of the western Zagros mts particularly of the mountain ranges around Karman and of the driest parts of Armenia and Azerbaijan.

Frietag (1977) gave the following "reasonable conclusions" of his investigations on the pollen record:

1) Annual precipitation : 150-135 mm 1/2-3/4 of the present day level ;
   1a) distribution : seasonal with same winter rain cycle as of today; and
2) temperature: may be lower than today but the drop in winter temp was probably more pronounced.

"This confirms only generally the earlier suggestion of Wright et al (1967) and Van Zeist (1967) that modern equivalents of the Pleniglacial vegetation should be looked for in northwest Iran or east Anatolia".

(2) Periods with predominant Pistacia woodland.

Early post-glacial: the semi-desert vegetation had an abrupt change into conditions typical of Pistacia (Amygdalus) woodland. In all, normal habitats did occur as open woodland communities *Pistacia, Acer monospessulanum, Amygdalus* and perhaps *Cerasus* and *Crataegus. Rheum, Ribes* flourished on rocky slopes and mesic Quercus braunii woodland community expanded in ravines and on northerly exposed slops. Their present day counterparts occur on the edge of *Q. braunii* area around the Zagros mts.

The change in climate from extremely semiarid to semi-arid state was slow and the mean annual precipitation could have been 250-300 mm, increasing gradually to 450-500 mm. Temperatures are not possible to derive from the available pollen record.

It may be mentioned here that Agrawal (1984) has stated that the climatic pattern of Kashmir follows a global trend: the warming up of the Pliocene, the glacial and interglacial oscillation of the Pleistocene etc. However, the Pleistocene cooling is stated to be " not abrupt but very gradual"

"More curious is the climatic data for 18,000 BP which globally is considered a glacial peak. In Kashmir we find the development of soil with organic debris derived from C3 type of plants and the broad leaved elements prominent in the pollen diagrams - all indicating climatic amelioration. Considering the evidence of mediterranean lakes (Perrot & Harrison in press) in the same latitude the Kashmir evidence seems to be in concordance".

Burbank & Johnson (1982) through fission track dating of volcanic ash met (with) in the valley have suggested a span of 4 m y for the origin of the Kashmir Karewas. Gupta et al (1983) on the basis of palynological investigations support Pliocene Age for a locality in Hirpur(karewa) Kashmir.

The comparative analysis of the stratigraphy of Central Asia and the northern Himalayan foothills is still disputable. However, there is reason to believe that upper Karewas is situated in the zone of normal polarity identified with lower Brunhes epoch; the lower Karewas correspond to the Matuyama to Gause palaeomagnetic epochs (Agrawal et al ,1979). The age of the loess mantle in the Kashmir valley is ca 100,000 yrs and "this is the time of the most intense loess accumulation of Centra Asia" (Dodonov, 1984).

A study of paleoenvironment of Hirpur in Kashmir (Gupta et al, 1984) has suggested the pre-quaternary age for the area with probability of Upper Pliocene. Good frequlencies of *Larix* from Hirpur sediments vis-a-vis its present day absence from the whole of the NW Himalaya poses an enigmtic problem. From the fossil records *Larix* had a wider distribution in Europe. This is attribbuted to climatic changes by Florin (1963)" Deterioration of climate and reduction in precipitation accopanied by strong winds were the main factors responsible for degeneradtion and consequent disappearance of *Larix*. Further they recored the following palaeo-climatic data on the basis of pollen assemblage zones and this reveals cyclic alterations:

HL-III . .1) Subtropical warm and dry phase: ***Pinus*** roxburghii

2) Warm temperature and wet: *Larix*[10]*-Quercus-Carya-Juglans- Engelhardtia-Fraxainus.*

3) Subtropical warm and moderately dry: *P. roxburghii-Juglans.*

4) Warm temperate and wet: *Larix-Quercus-Engelhardtia.*

5) Subtropical warm and dry: ***P. roxburghii***

6) Temperate and wet: ***Cedrus-Abies-Alnus***

7) Subtropical warm and dry: ***P.roxburghii***

8) Warm temperate and wet: ***Quercus-Cupressus-Larix-Carpinus-Alnus-Populus.***

9) a. subtrocipal warm and dry : *P.roxburghii- Celtis*-Oleaceae.

b. temperate and dry : *Abies-Picea-CedrusBetula-Corylus-Ulmus-Juglans*

c. subtropical warm dry: *P. roxburghii-Quercus-Viburnum*-Oleaceae.

10) Subtropical warm with progressively moist phase: *P.roxburghii- Quercus*-Oleaceae.

11) Warm temperate and less humid: *Quercus-Alnus-Engelhardtia*.

12) Temterate and wet: Liliaceae-ferns-*Abies.*

13) Cold temperate and moist: *Alnus- Quercus-Abies-Populus-Fraxinus.*

14) Cold temperate and moderately dry: *Abies-Picea-Cedrus-Betula-Carpinus-Alnus-Juglans.*

The other localities (Karewas) investigated on the basis of palynostratigraphy in Kashmir show a more or less similar climatic inference as revealed by broad vegetational development.

---

10. Presence of *Larix* needs re-examination. The arguments given are not tenable.

3

BOTANICAL REGIONS

CENTRAL ASIA

Asia is divisible into six major physical divisions: the central föld mountainous innermost plateau (including Iran (W). Afghanistan and Baluchistan (E) very dry with extreme climate: hot summers and cold winters): the eastern volcanic islands; the northeast highlands ; the southern table-lands; the alluvial plains of southeast Asia ; and the northern plain (Stenbridge, 1956). Of these the central fold mountains and plateau are pertinent here. These form a climatic and physical barrier between southeast Asia and northwest Asia; and are normally distinguished into six ranges and plateaus (Fig.10):1) the Armenian Knot, 2) plateau of Asia Minor bounded by Pontic mountains, Anti-Taurus and Taurus , 3) plateau of Iran in the bossom of Elbruz mts, Hindukush, Kurdistan highlands and Sulaiman mts,4) Tarim basin bounded by Tien Shan and Kun Lun, 5) the plateau of Tibet margined by Kun Lun, and 6) the Himalaya.

Winter temperatures far below the freezing points occur in central and northern Asia,and owing to their this "intense cold a region of high pressure with overflowing winds is found over Central Asia". The summer temperatures in Central Asia go up to 80^0 F. Most of the Central Asia has poor steppe and desert vegetation. However in the moutain belt it varies with elevation. Thus the major part is of Arid lands. (In Iran the arid land is nearly half sand and deserts, rest poor steppe,the irrigated areas are rich in crops. Greater part of Afghanistan is stony slopes,arid plateaus are with stretches of rough herbage,the irrigated valleys and plains are few").

The Chinese Central Asia includes two regions: Sinkiang and Tibet (now annexed by the Chinese). Sinkiang is bounded by Altai mts and the Kun Lun. Much of its area is sand dunes and poor steppes. Irrigated areas are cultivated. Mongolia is largely a desert (Gobi desert) but receives enough rains on its margins. Tibet lies between Kun Lun and the Himalaya,has severe climate with a vast barren desert and scanty pasture land.

In Tibet plateau (Fig.11) (elv 4000 m,area ca 2,201,000 km^2) animal husbandry has been the main agricultural activity for centuries and the land is rich in pastures. The humidity falls gradually from southeast to northwest but there is little difference in thermal conditions between the southern and northern parts at the same elevation. Tibet has been divided into ten natural regions by Sun (1983) as follows :

1. humid tropical evergreen rainforest with latosols:
2. sub-humid warm temperate xerophilous forest of shrubland with drab soils;
3. subhumid temperate coniferous forest with burozems;
4. subhumid cold temperate alpine meadow with felty sod soils;
5. semi-arid temperate shrubby steppe with shrub by steppe soils;
6. semi-arid cold temperate alpine steppe with alpine steppe soils;
7. semiarid frigid alpine permafrost desert steppe with cold desert steppe soils;
8. arid temperate desert shrubby ssteppe; with surface accumulated carbonate shrubby steppe soils;
9. arid cold temperate alpine desert with cool desert soils; and
10. arid frigid alpine permafrost desert with cold desert soils.

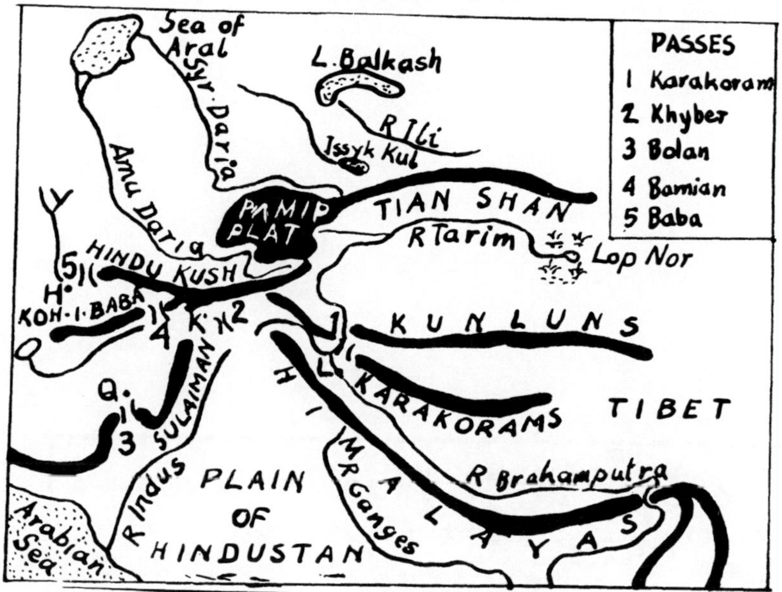

Fig. 10 (cf. Stenbridge, 1956) : Central Asian mountains.

Fig. 10 b: Central Asian Region and adjoining phytogeographic regions.

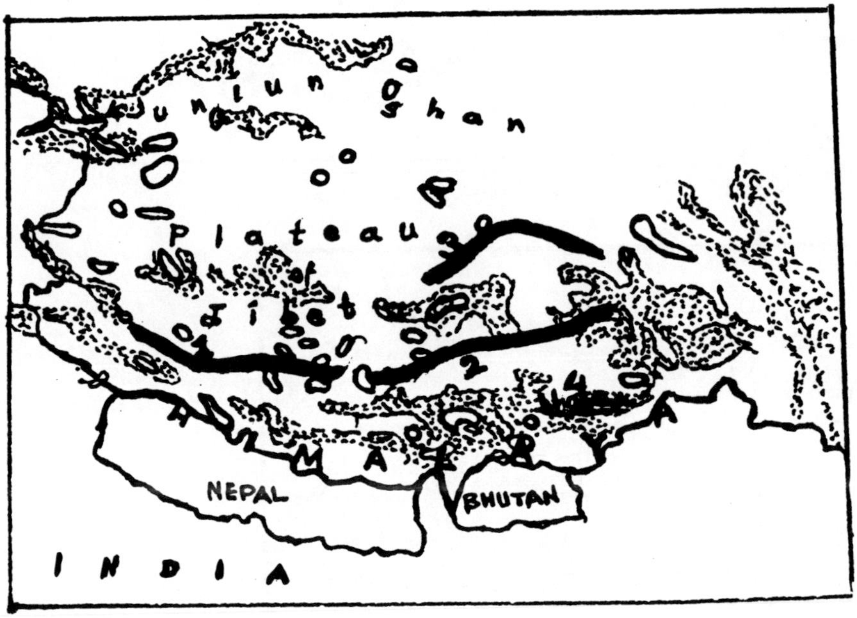

Fig. 11 . Tibet (physical): 1 Kailash Range, 2 Nyenchen-tanglha Range, 3 Tangla Range, 4 Lhasa.

The mean altitude is over 16,500 ft. the valleys lying at 14,500-17,500 ft. The temperatures are severe 15-16°F in summer day and zero or lower at night; in winter temp touches -40° F or more. Vegetation is rare. However herbaceous plant cover occurs in spaces between slopes, few forests exist near Lhasa consisting of shrubs, bushes, juniper and tamarisk. A few other trees found are pine, fir, willow. holly tree, birch, cedar, oak and elm. The main crop is barley. Pastures are luxuriant, sheep and yak are valuable herd.

Botanico-geographical regions of Central Asia

Grubov (1963) described the botanico-geographical regions of Central Asia, dividing it into three major provinces: 1) Mongolia, 2) Djungaro -Turanica, and 3) Tibetica (Fig.12). Each province was further subdivided into floristic areas, enumerating typical plant species in each.

I. Provincia Mongolica

It was divided into three major sub-provinces: 1) Mongolia, 2) Kaschgarica, and 3) Tsaidam. The typical species recorded are:

Aconitum gynandrum, Adenophora hymalayana, Alhagi sparsifolia, Allium mongolicum, A. polyrrhizum,Ammopiptanthus nanus,Anabasis aphylla,A.brevifolia, Artemisia arenaria, A.frigida, Asterothamnus centraliasiaticus. A. fruticosus, Brachanthemum gobicum, Anthrophytum longibracteatum,Caryopteris mongolica, Calligonum zaidamense, Crepis flexuosa, Dilophia salsa, Dracocephalum tanguticum, Elymus dasystachys, Ephedra przewalskii, Eurotia ceratoides, E.compacta, Geranium pylzovianum, Gymniocarpos przewalskii, Haloxylon ammodendron, Haplophyllum dahuricum, Hypolitia achilleoides, H. gobica, H. trifida, Iljinia regelii, Juniperus zaidamensis, Kalidium gracile, Karelinia caspica, Lagotis brachystachya, Ligularia sagitata, Lomonium aureum, L.roborowskii, Lycium ruthenicum, Microulia sikkimensis, Myricaria dahurica, Nitraria sibirica, N. sphaerocarpa, N.tangutorum, Oxytropis tragacanthoides, Potaninia mongolica, Piptanthus mongolicus, P.nanus, Pappagrostis pappophroea, Poacynum handersonii, Psammochloa billosa, Reaumuria kaschgaria, R.soongorica, R. trigyna, Rheum spiciforme, Roborowskia mira, Rumex nepalensis, Salsola arbuscula, S. abratanoides, S.passerina, S. zaidanica, Saussurea sorocephala, Salvia roborowskii, Stipa glareosa, S.gobica, Tanacetum brachanthemoides, T.leaschgaricum, Sympegma regelii, Tetraena mongolica, Thermopsis lanceolata, Thylacospermum rupifragum, Trachomitrium lancifolium, Zygophyllum kaschgarica, Z. rosovii, Z. xanthoxylon.

I A Mongolia

IA (1) Chobdo : *Anabasis brevifolia, Arnebia decumbens, Asperula humifusa, Carex soongorica, Chenopodium frutescens, Eremostachya molucelloides, Londesia eriantha, Serratula altavica, Stipa glareosa.*

IA (2) Altai mongolicus : *Aconitum altaicum, Artemisia altaiensis, Astragalus altaicus, A. schangimanus, A. sphaerocystis, Biebersteinia odora, Campanula altaica, Chorispora bungeana, Deschampsia koelerioides, Dianthus hoeltzeri, Draba altaica, Echinops intergrifolius, Euphorbia alpina, Ferula teterrima, Gentiana turkestanicum, Oxytropis altaica,O.ladyginii, O.martzanovii, Pedicularis altaica, Potentilla imbricata, Sanguisorba alpina, Scaligera setacea, Trollius altaicus.*

IA (3) Cischingania : *Aconitum komarovii, Agropyron turczaninovii, Anemarrhena asphodeloides. Betula manschurica, Cirsiun vlassovianum, Elymus chinensis, E. dahuricus, E.sibiricus, Filifolium sibiricum, Hemerocallis flava, Iris dichotoma, Koeleria gracilis, Malus baccata, Padus asiatica, Paeonia lactiflora, Patrinia scabiosifolia, Platycodon grandiflorus, Poa sphondilodes, Rhamnus ussuriensis. Stipa baicalensis, S.grandis, S.sibirica, Synurus deltoides, Ulmus propinqua.*

IA (4) Chalcha media: *Agropyron cristatum, Allium anisopodium, A.polyrrhizum, Anabasis brevifolia, Artemisia adamsii, Arenaria frigida, Atriplex* sps, *Caragana microphylla, Carex duriuscula, Cleistogenes squarrosa, Cymbaria dahurica, Delphinium cheilanthum, Elymus chinensis, E. dasystachya (= Aneurolepidium pseudagropyrum), Euphorbia discolor, Iris biglumis. Kalidium graciale, Koeleria gracilis, Lilium tenuifolium, Peucedanum baicalense, Polygonum divaricatum, Reaumuria soongorica, Ribes pulchellum, Salsola passerina, Scutellaria sordifolia, Silene jenisseensis, Spiraea flexuosa, Stipa glareosa, S.gobica, S.grandis, Suaeda* sps., *S.krylovii (= decipiens).*

IA (5) Mongolia orientalis : *Acer ginnala, Agriophyllum arenarium, Allium mongolicum, A.odorum, A.polyrrhizum, Armeniaca Sibirica, Alnus sibirica, Artemisia holodendron, A.ordosica, Betula dahurica, B.manschurica, B.platyphylla, Biota orientalis, Calamogrostis epigeios,*

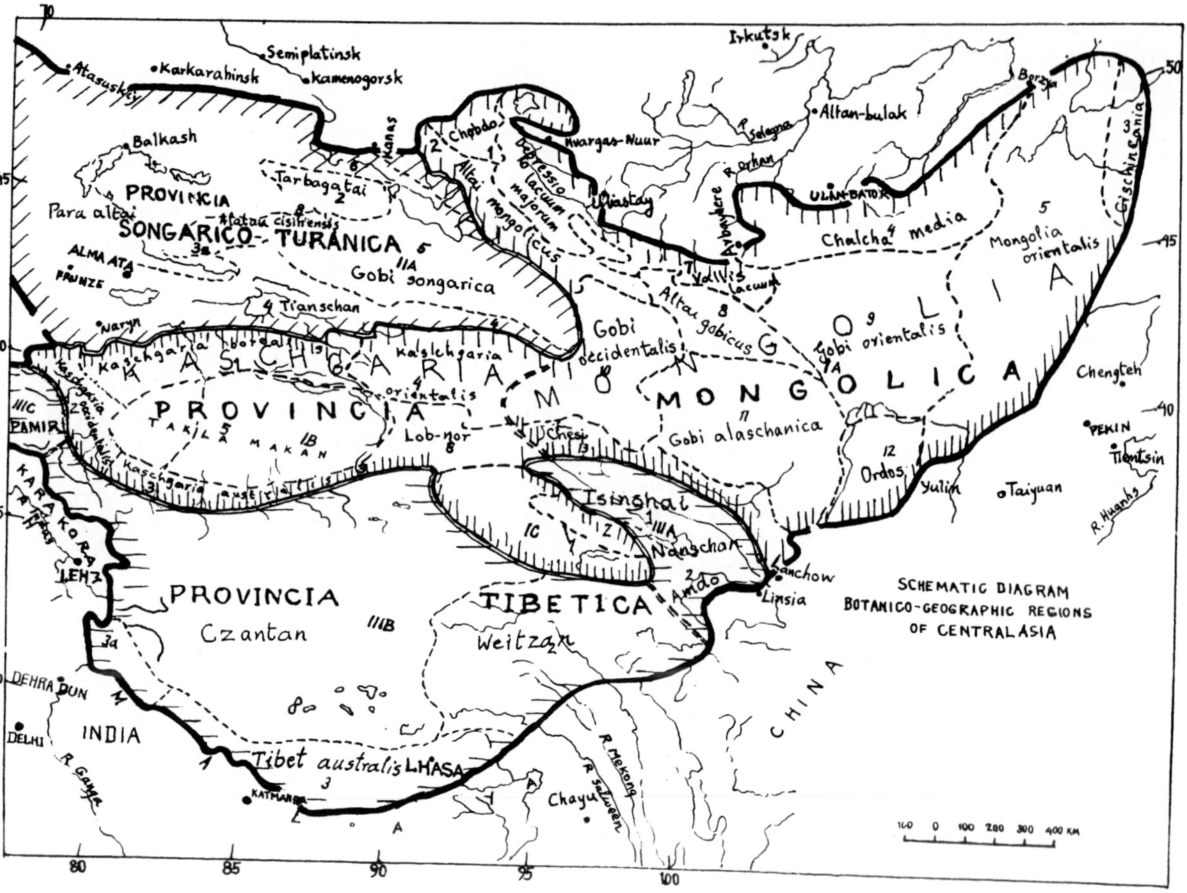

Fig. 12 : Botanico-geographic regions of Central Asia (cf Grubov, 1963).

Caragana microphylla, Cleistogenes squarrosa, Cornus macrophylla, Crataegus dahuria, Elymus chinensis, Euonimus maackii, Festuca sulcata, Hemerocallis minor, Kalidium foliatum, K.grciale, Kochia gracilis, Lilium tenuifolium, Malus baccata, Melica scabra, M. virgata, Ostryopsis davidiana, Padus asiatica, Paeonia lactiflora, Pinus silvestris, Poa attentuata, Psamochloa villosa, Populus laurifolia, P.tremula, Pugionium cornutum, Quercus mongolica, Reaumuria soongorica, Rhamnus arguta, Salix mongolica, S.pubescens, Spiraea hailarensis, S.trilobata, Sorbus pohuashanensis, Stipa krylovii, S. sibirica, Thalictrum squarrosum, Tilia mongolica,ulmus propinqua, U. pumila, Viburnum sargentii, Vicia unijuga.

IA (6) Depressio lacuum majorum : *Aaluropus litoralis, Agriophyllum arenarium, Agropyron cristatum, Allium polyrrhizum, Anabasis brevifolia, Artemisia caespitosa, A.frigida, A.maritima, A xerophytica, Asterothamnus centraliasiaticus, Betula microphylla, Chenopodium frutescens, Cleistogenes squarrosa, Caragana leucophloa, C. bungei, C. spinosa, Ceratocarpus arenarius, Elymus giganteus, Eurotia ceratoides, Helimodendron halodendron, Hedysarum mongolicum, Kalideum foliatum, K. gracile, Lycium ruthenicum, Nanophyton crinaceum, Populus laurifolia, P.pilosa. Psammochloa villosa, Pugionium pterocarpum, Reaumuria soongorica. Salsola abratanoides, Stipa krylovii, S. glareosa, Tanacetum achilleoides.*

IA (7) Vallis lacuum : *Allium polyrrhizum, Anabasis brevifolia, Artemisia arenaia, A.frigida, Cleistogenes mutica, Caragana bungei, C. pygmaea, Eurotia ceratoides, Hedysarum mongolicum, Hyppolitia achilleoides, Kalidium gracile, Psammochloa villosa, Reaumuria soongorica, Salsola abrotanoides, S. arbuscula, S.posserina, Stipa glareosa, S.gobica, S.krylovii.*

IA (8) Altai gobicus : *Allium mongolicum, A. polyrrhizum, Agropyron cristatum, Amygdalus mongolica, A.pedunculata, Anabasis brevifolia, Artemisia arenaria, A. disjuncta, A. frigida, A. pamirica, A.xerophytica, Asterothamnus centraliasiaticum, Betula microphylla, Caragana bungei, C. leucophloea, Carex pseudofoetida, Clematis fruticosa, C. tangutica, Cotoneaster melanocarpa, C. mongolica, C. uniflora, Ephedra przewalskii, Eurotia ceratoides, Grossularia acicularis, Hedysarum mongolicum, Juniperus sabina, Kalidium foliatum, K. gracile, Kochia prostrata,Lonicera altaica, L. macrophylla, Nitraria sphaerocarpa, Oxytropis aciphylla. Populus densa, P. pilosa, Potaninia mongolica, Psammochloa villosa, Reaumuria soongorica, Ribes rubrum, Rosa acicularis, Salsola arbuscula, S. abrotanoides, S. passerina, Stipa glareosa, S.gobica, S. krylovii.*

IA (9) Gobi orientalis : *Anabasis brevifolia, Artemisia xanthochroa, A. xerophytica, Brachanthemum gobicum, Caragana bungei, Eurotia ceratoides, Kalidium gracile, Polaninia mongolica, Reaumuria soongorica, Salsola passerina.*

IA (10) Gobi occidentalis : *Alhagi sparsifolia, Anabasis brevifolia, Calligonum mongolicum, Caragana leucophloea, Elaegnus moorcroftii, Ephedra przewalskii, Glycyrrhiza uralensis, Iljinia regalii, Lasiogrostis splendens, Lycium ruthenicum, Nitraria sibirica, Picea asperata, Phragmites communis, Poacynum hendersonii, Populus diversifolia, Reaumuria soongorica, Rhamnus* sps, *Salix caspica, Sophora alopecuroides, Stipa gobica,Sympegma regelii, Tamarix ramosissima, Zygophyllum leaschganicum, Z. xanthoxylon.*

IA (11) Gobi alaschanica : *Agriophyllum arenarium, Ammopiptanthus mongolicus, amygdalus mongolicus, Anabasis brevifolia, Artemisia ordosica, A. sphaerocephala, Atraphaxis frutescens, Astragalus discolor, Caragana bungei, C.brachypoda, C. microphylla, C.stenophylla. Convolvulus gortschakovii, C.tragacanthoides, Gymnocarpos przewalskii, Hedysarum mongolicum, H scoparium, Nitraria sphaerocarpa, N.sibirica, Picea asperata, Pinus tubulaeformis, Potaninia mongolica, Psammochloa villosa, Ptilagrostis pelliotii, Pugionium cornutum, P. dolabratum, Raumuria soongorica, Salsola passerina, S. liricifolia, Stipaglareosa, S. gobica, Sympegma regelii, Tamarix ramosissimum, Zygophyllum xanthoxylon.*

IA (12) Ordos : *Agriophyllum arenarium, Ammopiptanthus mongolicus, Artemisia arenaria, A. frigida, A. ordosica, A. sphaerocephala, Atraphxis frutescens, Caragana boschinskyui, C. microphylla, Convolvulus tragacanthoides, Corispermum* sps, *Hedysarum mongolicum, H. scoparium, Kalidium foliatum, K. gracile, Lasiagrostis splendens, Nitraria sibirica, Psammochloa villosa, Potaninia mongolica, Reaumuria* *soongorica, R. trigyna, Salsola passerina, Tetraena mongolica, Zygophyllum xanthoxylon.*

IA (13) Chesi : *Anabasis brevifolia, Brachanthemum neuschanicum, Cancrinia maximowiczii, Kalidium cuspidatum, K. gracile, Reaumuria soongorica, Salsola passerina, Stipa gobica, Zygophyllym xanthoxylon.*

IB Kaschgaria

Achiropus litiralis, Calamogrostis pseudophragmites, Clematis orientalis, Cynanchum pubescens, Elaegnus moorcroftii, Glycyrhiza glabra, G. inflata, Helimodendron halodendron, Halostachya cuspica, Hippophae rhamnoides, Lasiagrostis splendens, Lycium ruthenicum, Karelinia caspica, Nitraria roborowskii, N. sibirica, Phragmites communis, Picea schrenkiana, Poacynum hendersonii, Populus diversifolia, P. pruinosa, Sophora alopecuroides, Tamarix laxa, T. ramosissimma, Trachomitrium lancifolium.

IB (1) Kaschgaria borealis : *Alhagi sparsifolia, Anabasis aphylla, A. truncata, Artemisia arenaria, A. frigida, A.ksaschagarica, Arthrophytum longibracteatum, Calligonum roborowskii, Caraganajubuta, Ephedraprzewlskii, Festuca sulcata, Gymnocarpos przwalskii, Iljinia regelii, Juniperus pseudosabina, Kalidium cuspidatum, Larelinia caspica, Leontopodium ochroleum, Nitraria sphaerocarpa, Oxytropis* sps, *Reaumuria soongorica, R. kaschgarica, Salsola junatovii, Sympegma regelii, Tanacetum brachanthemoides, T. kaschgaricum, Zygophyllum kaschgaricum.*

IB (2) Kaschgaria occidentalis : *Ammopiptanthus nanus, Artemisia parvula, A. rhodantha, Cobresia capilliformis, Festuca sulcata, Helimodendron halodendron, Leucopoa olgae, Lonicera semenovii, Populus diversifolia, P. pruinosa, Roborowskia mira, Rosa kaschgarica, Salix tenuifolia, Sorbus tianschanica, Spiraea* sps, *Stipa glareosa.*

IB (3) Kaschgaria australis : *Acantholimon diapensioides, Androsace squarrosula, Artemisia parvula (?), A.rhodantha, Asterothamnus fruticosus, Berberis nummularia, Calligonum roborowskii, Ephedra przewalskii, Eurotia ceratoides, Gymnocarpos przewalskii, Kalidium cuspidatum, Myricaria alopecuroides, M. elegans, Nitraria sphaerocarpa, Oxytropis poncinsii, Populus laurifolia, Ptilagrostis purpurea, Reaumuria kaschgarica, R. trigyna, R. soongorica, Rosa kaschgarica, Sibbaldianthe tetranda, Stipa breviflora, S. caucasica, S.glareosa, Sympegma regelii, Tanacetum tibeticum, T. xylorrhizum, Thylacospermum rupifragum, Ulmus pumila, Zygophyllum kaschgaricum.*

IB (4) Kaschgaria orientalis : *Alhagi sparsifolis, Anabasis brevifolia, Ephedra przewalskii,Iljnia regelii, Kalidium gracile, Nitraria, sphaerocarpa, Reaumuria soongorica, Sympegma regelii, Tanacetun brachanthemoides.*

IB (5) Takla-Makan : *Aristida pennata.*

IB (6) Lob-nor : *Halostachys caspica, Haloxylon ammodendron, Kalidium gracile, Poacynum hendersonii, Trachomitrium lancifolium.*

IC TSAIDAM

IC (1) Tasaidam planitierus

IC (1 a) boreali- occidentalis : *Artemisia xanthochroa A. xylorrhiza (?) , A. traphaxis ,Calligonum zaidamense , Carex* sps. *Clematis tangutica, Elymus dasystachys,Ephedra glauca, Gymnocarpos przewalskii, Halogeton glomeratus , Hedysarum arbuscula, Juncus atrofuscus, Kalidium gracile, Limonium aureum, Oxytropis aciphylla ,O.glabra, Psammochloa villosa , Salsola* sps *.S. arbuscula, Saussurea salsa, Tanacetum falcatolobatum, Triglochin palustris.*

IC (1 b) australis : *Allium* sps, *Aster altaicus, Artemisia anthifolia, A. macrocephala., A. pectinata, A. scoparia, A. subdigitata, A. xanthochroe, Asterothamnus centraliasiaticus, Astragalus monophyllos, Bolboschoenus affinis, Calamogrostis macilenta, C. pseudophragmites, Carex aterrima, C.caesposa, C. orbicularis, Clematis orientalis, C. tangutica, Comarum salesovianum, Cynomorium soongaricum, Dasiphora fruticosa, D. parvifolia (?), Ephedra przewalskii, Elymus dasystachys, Eurotia ceratoides, Gentiana squarrosa, Glaux maritima,Glycyrrhiza uralensis, Hedysarum multijugum, Heleocharis intersita, H. purpurea, Heleropestes salsuginosa, Hippuris vulgaris, Juniperus pseudosabina, Juncus atrofuscus, Kalidium cuspidatum, Lycium chinense, L. turcomanicum, L. ruthenicum, Myricaria alopecuroides, M. squamata, Nitraria langutorum, N. sibirica, Oxytropis falcata, Poacynum hendersonii, Polygonum sibiricum, Populus diversifolia, Potamogeton pectinatus, Puccinellia sps, Ranun-*

culus tanguticus, Raeumulria soongorica, R. kashagarica varprzewlskii, Saussurea salsa, Scirpus disconcavus, S. validus, Sympegma regelii, Thermopsis lanceolata, Trichomitrium lancifolium, Triglochin maritima, T. palustris, Trichophorum pumilum, Typha laxmanii, T. minima, Utricularia vulgaris, Youngia paleacea.

IC (1 c) Sirtyn : *Agropyrum cristatum, Allium strictum, A. tenuissimum, Artemisia pectinata, Astragalus monophyllos, Cobresia royleana, Comarum salesovianum, Dasiphora fruticosa, Eurotia ceratoides, Glaux maritima, Hedysarum multijugum, Hordeum brevisubulatum, Iris biglumis, Kalidium cuspidatum, K.gracile, Limonium aurcum, Lomatogonium rotatum, Mulgedium tataricum, Oxytropis tragacanthoides, Polygonum sibiricum, Potentilla dealbata, P. moorcroftii, Ranunculus tanguticus, Raeumuria kaschgarica* var *nanschanica, R. trigyna, R. soongarica. Salsola abrotanoides, Saussurea salsa, Stipa breviflora (?), S. glareosa (?), Sympegma regelii, Tanacetum falcatolobatum.*

IC (2) Tsaidam montanus

Aconitum gymnandrum, Adenophora hymalayana, Agropyron cristatum, A. desertorum, Amblyotropis stenophylla, Androsace erecta, Artemisia frigida. A. pectinata, A. sieversiana, Atragalus capillipes, A. monophyllos, A. tataricus, Caragan jubata, C. densa, C. roborowskii, Cleistogenes mutica, Clematis tangutica, Comarum selesovianum, Crepis flexuosa, Dasiphora fruticosa, Dilophia salsa, Dracocephalum tanguticum, Eurotia ceratoides, E. compacta, Festuca sulcata, Gentiana aperta, G. straminea, Geranium pylzovianum, Hippophae rhamnoides, Juniperus squamata, J. zaidamense, Kalidium cuspidatum, K. graciale, Lagotis brachystachya, Ligularia sagitata, Microula sikkimensis, M. tibetica, Morina chinensis, Oxytropis imbricata, O. ochrocephala, O.kansuensis, O. tragacanthoides, Pedicularis kansuensis, Picea asperata, Poa kakonorica, Polygonum statice, P. viviparum, Potentilla bifurca, Reaumuria soongorica, R. trigyna, Rheum pumilum, R. spiciforme, Rumex nepalensis, Salsola abrotanoides. Saussurea semifasciata, S. sorocephala, Salvia roborowskii. Stipa breviflora, S. bungeana, S. inbrians, S. penicellata, Sympegma regelii, Thylacospermum rupifragum.

II. Provincia Djungaro- Turanica

Alhagi kirghisorum, A. sparsifolia, Allium polyrrhizum, Ammopiptanthus mongolicus, A. nanus, Anabasis aphylla, A. brevifolia, A. salsa, Artemisia frigida, Artiplex cana, Borszezowia aralocaspica, Clematis soongorica, Ephedra przewalskii, Elvmus junceus, Girgensohnia opposiflora, Gymnocarpos przewalskii, Halimocneis karelinii, Halimodendron halodendron, Haloxylon ammodendron, H. persicum, Iljinia regelii, Kalidium caspicum, Lasiagrostis caragana, L. splendens, Lycium ruthenicum, Nanophyton arinaceum, Niedzewdzkia semiretschenskia, Nitraria sphaerocarpa, Poa bulbosa, Poacynum hendersonii, P. pictum, Potaninia mongolica, Reaumuria soongorica, Spiraenthus schrenkianus, Salsola arbusculiformis, S. arbuscula, S. orientalis, S. lessingiana, S. gobica, S. sareptana, Suaeda dendroides, S. physophora. Sympegma regelii, Tatraena mongolica, Timouria saposhikovii, Zygophyllum xanthoxylon.

II A Djungaria: *Anabasis aphylla, A. brevifolia, A. salsa, Artemisia borotalensis, A. kaschgarica, A. terrae-albae, Atraphrxis frutescens, Cleistogenes thorldii, Convolvulus fruticosus, Eurotia ceratoides, Haloxylon ammodendron, Kochia prostrata, Ptillagrostis pelliotii, Salsola arbuscula, Stipa glareosa, Sympegma regelii.*

IIA (1) Para - Altai: *Aconitum excelsum, Agropyron desertorum Amygdalis nana, Anabasis brevifolia, Anemone altaica, Archangelica decurrens, Artemisia frigida, A. kaschgarica, A. maritima, A. semiarida, A. sublessingiana, Betula pubescens, Bupleurm aureum, Cacalia hastata, Caragana pygmaea, Cleistogens squarrosa, C. thoroldii, Cotoneaster melanocarpa, Delphinium* sps. *Festuca sulcata, Koeleria gracilis, Larix sibirica, Ligularia altaica, Lilium martagon, Lonicera altaica, L. hispida, L. microphylla, L. tatarica, Paeonia anomala, Picea obovata, Rosa acicularis, R. alberti, R. spinosissima, Spiraea chamaedrifolia, S. hypericifolia, S. media, Stipa capillata, S. glareosa, S. gobica, S. lassingiana, S. orientalis, S. rubens, S. sibirica, S. sareptana, Trollius altaicus, Veratrum lobelianum.*

IIA (2) Tarbagatai : *Aconitum excelsum, A. soongoricum, Agroyron abolinii, A. czimganicum, A, desertorum A. repens A. schtenkianum, A. tianschanicum, A. tarbataicum, Alchemilla cryp-*

topleura, Anthoxanthum odoratum, Allium bidentatum, A. polyrrhizum, A. hyminorrhizum, Alopeucurus soongoricus, A. pratense, Amygdalus ledebouriana, Anemone schrenkiana, Aster alpinus, Aquilegia glandulosa, Astragalus arbuscula, Bromus inermis, Bupleurum aureum, Atapohaxis laetevirens, A. virgata, Calophaca howenii, Cerasus tianschanica. Caragana frutex, C. leucophloea, Cobressia bellardii, C. smironovii, Corallorhiza innata, Cotoneaster melanocarpa, crepis sibirica, Dactylis glomerata, Daphne altaica, Deschampsia caespitosa, Dracocephalum altainse, D.nutans, Ephedra distachya, Eremurus altaicus, Euonymus semenovii, Festuca rubra, F. sulcata, Ferula soongorica, Hedysarum neglectum, Helictotrichon asiaticum, H. pubescens, H. tianschaNicum, Heracleum dissectum, Hordeum turkestanicum, Ixiolirion tataricum, Juniperus pseudosabina, J. sibirica, Kochia prostrata, Lathyrus gmalinii, Ligularia altaica, Lilium martagon, Lonicera tatarica, Moneses uniflora, Minuartia biflora, Origanum vulgare, Papaver croceum, Phleum alpinum, P. phleoides, P. pratensis, Picea schrenkiana, Poa alpina, P. bulbosa, P.dschungarica, P. .nemoralis, P. pratensis, P. relaxa, P.sibirica, Polygonum alpinum, P. nitens, Pyrethrum achillefolium, Pyrola grandiflora, P.minor, Rhamnus Cathalrtca, Ribes meyeri, Rosa acicularis, R. beggariana, R. spinosissima, Saussurea frolovii, Sanguisorba alpina, Serratula marginata, Spiraea hypericifolia, Stipa capillata, S. lessingiana, S. rubens, Schultzia crinita, Swertia obtusa, Trisetum virescens, Tulipa schrenkii, T. biflora, Trillius altaicus, Trisetum sibiricum.

IIA(3) Alatau cisiliensis*: Amygdalus ulmifolia, Andrachne rotundifolia, Armeniaca vulgaris, Artemisia frigida, A. sublessingiana, Aster alpinus, Astragalus tibeticus, Berberis heteropoda, Brachypodium pinnatum, Bromus inermis, Calamogrostis epigeios, Carex turkestanica, C. stenocarpa, Celtis caucasica, Cleistogenes squarrosa, C. thoroldii, Cobresia capilliformis, C. pamiroalaica, Crambe kotschyana, Dactylis glomerata, Cotoneaster melanocarpa, C. multifida, Festuca sulcata, Gentiana falcata, Helictrotrichon asiaticum, H. tianschanicum ,Inula rhizocephala, Kochia prostrata, Koeleria gracilis, Leontopodium ochroleucum, Lonicera altmanii, L. hispida, L. karelinii, L. microphylla, L. sterantha, Malus sieversii, Nigella integrifolia, Orthurus heterocarpus, Oxytropis* sps, *Phleum phleoides, Poa angustifolia P. bulbosa, Potentilla nervosa, Primula algida, Rubus idaeus, Spiraea hypericifolia, Stipa capillata, S. glareosa, S. gobica, S. lessingiana, S. kirghisorum, S. orientalis, S. rubens, S. sareptans, Thuidium abientinum, Valeriana ficariifolia.*

IIA(4) Tien Schan *:Acantholimon tianschanicum, Aegopodium alpestre, Aellenia glauca, Agropyron cristatum, A. tianschanicum, Allium polyrrhizum, Alopecurus pratensis, Anabasis brevifolia , Andropogon ischaemun, Arenaria formosa, Armeniaca vulgaris, Artemisia austriaca, A. borotalensis, A. frigida, A. kaschgarica, A. santolinifolia, A. terrae-albae, Aster alpina , Astragalus tibetanus, Berberis heteropoda, Betula microphylla, B.tianschanica, B.verrucosa, Brachypodium pinnatum, Bromus inermis, Calamogrostis langsdorfi, C. obtusata, Caragana acanthophylla, C.dasyphylla, C. laeta, C.leucophloea, Carex stenocarpa, C. tianschanica, C.turkestanica, Cerastium dahuricum, Cleistogenes thoroldii, Clematis sibirica, Cobresia capilliformis, C.pamiroalaica, Corallorhiza innata, Cotoneaster malanocarpa, Crataegus altaica, C.soongorica, Crepis sibirica, Dactylis glomerata, D.spicatum, D.trisetum, Dasiphora fruticosa, Doronicum altaicum, Draba alpina, Dryanthe tetranda, Entodon orthrocarpus, Ephedra sinica, Euonymus semenovii, Eurotia ceratoides, Festuca kryloviana, F. rubra, F. sulcata, Goodyera repens, Helictotrichon pubescens, H. schellianum, Hylocomium proliferum, Hypnum revolutum, Inula rhizocephala, Juniperus sabina, J.turkestanica, Kochia prostrata, Koeleria gracilis, Lagotia glauca, Lasiagrostis caragana, Leontopodium ochroleucum, Lonicera coerulea, L. hispida, L. karelinii, L. microphylla, Moneses uniflora, Malus sieversii, Mulgedium azureum, Oxytropis* sps, *Papaver nudicaule, Phleum pratense, Picea schrenkiana, Poa pratensis, P. numeralis P.stepposa, Populus laurifolia, Ptilogrostis purpurea, P.subsessiliflora, Potentilla nervosa, Primula algida, Pyrola minor, P. rotundifolia, Rhytidiadelpus triquetrus, Ramischia secunda, Rosa alberti, R. beggariana, R. platyacantha, Salix hastata, S. iliensis, Saxifraga hirculus, Senecio soongoricus, Solidago virga-aurea, Spiraea hyopercifolia, S. tianschania, Stipa breviflora, S. capillata, S. glareosa, S. orientalis, S. lassingiana, S. kirghisorum, S. rubens, S.sereptana, Sorbus* *tianschanica, Tamarix ramosissima, Thuridium abietinum, Thylacospermum caespitosum, Trisetum spicatum, Ulmus pumila, Veronica spicata.*

IIA(5) Gobi djungarica: *Agriophyllum arenarium, Allium polyrrhizum, Alyssum desertorum, Anabasis aphylla, A. brevifolia, A. eriopoda, A. salsa, A. truncata, Artemisia borotalensis, A. santolinifolia, A. santolina, A. terrae-albae, Astrapohaxis virgata, Aristida arbuscula, A. pennata, Atriplex cana, Betula microphylla, Brachathemum mongolicum, Calligonum mongolicum, Caragana leucophloea, Chondrilla lejesperma, Cleistogenes thoroldii, Cotoneaster melanocarpa, Ephedra przewalskii, Eremopyrum orientale, Eremostachya molucelloides, Eremurus inderiensis, Euphorbia rapulum, Eurotia ceratoides, E. eversmanniana, Gagea filiformis, Glycyrrhiza glabra, Grossularia acicularis, Gymnocarpos przewalskii, Halogeton glomeratus, Haloxylon ammodendron, H. persicum, Halocnemum strobilaceum, Helimodendron halodendron, Iljinia regelii, Ixilirion tataricum, Juniperus pseudosabina, Kalidium caspicum, Karelinia caspica, Kaschgaria lomorovii, Larix sibirica, Lepidium perfoliatum, Leptaleum filifolium, Limonium otolepis, L. suffruticosum, Leontice incerta, Lonicera microphylla, Lycium ruthenicum, L. turcoimanicum, Malcomia africana, Nanophyton erinaceum, Nitraria sibirica, Oxytropis aciphylla, Poa bulbosa, Poacynum hendersonii, Populus laurifolia, Ptilagrostis pelliotii, Pyrethrum discoideum, Reaumuria soongorica, Ribes rubrum, Rosa laxa, Saussurea salsa, Schismus arabicus, Salsola affinis, S. arbuscula, S. bracheata, S. ferganica, S. laricifolia, S. korschinskii, S. nitrata, S. praecox, S. rigida, Spiraea hypercifolia, Spirorhynchus saleuosus, Stipa breviflora, S. glareosa, S. gobica, S. orientalis, Suaeda dendroides, S. physophora, Sympegma regelii, Tamarix hispida, T. laxa, T. ramosissimum, Trachomitrium lancifolium, Trigonella arcuata, Tulipa biflora, T. ileinsis, Zygophyllum xanthoxylon, Tetracme quadricornis, T. recurva.*

IIA (6) Depressio Zaissan: *Anabasis aphylla, A. salsa, Artemisia arenaria, A. gracilescens A. terra-albae* var *massagetovii, Atriplex cana, Calligonum rubicundum, Caragana frutex, Chondrilla pauciflora, Elymus giganteus, Eremostachya molucelloides, Eurotia ceratoides, Festuca sulcata, Halimodendron halodendron, Halocnemum strobilaceum, Haloxylon ammodendron, H. persicum, Kalidium, Ligularia songorica, Megacarpaea megalocarpa. Populus, alba, P. nigra, Sophora alopecuroides, Spiraea hypercifolia, Stipa sareptana, Tulipa biflora.*

IIA (7) Depressio Dsharkent : *Allium lasiophyllum, Ammodendron argenteum, Anabasis elatior, A. eriopoda, A. salsa. Arnebia guttata, Aristida penntata, Artemisia santolina* var *suaveolens, A. terraealbae, Arthrophytum balschanschense, A. iliense, A. longibracteatum, Astragalus commixtus, Berberis iliensis, Carex pachystilis, C. physodes, Colpodium humile, Cleistogenes thoroldii, Convolvulus fruticosus, Cousinia tenella, Elaegnus oxycarpa, Eremostachya molucelloides, Euphorbia rapulum, Fraxinus potamophila, Eurotia ceratoides, Goldbachia laevigata, Halimocnemis semenovii, Haloxylon ammodendron, Haplophyllum multicaule, Horaninowia minor, Iijinia regelii, Koelpinia linearis, Leptaleum filifolium, Lepidium perfoliatum, Malcolmia scorpioides, Matricaria subglobosa, Meniocus linifolium, Nanophyton erinaceum, Nonnea caspica, Papaver pavoninum, Poa bulbosa, Populus alba, P. diversifolius, P. litwinoviana, Reaumuria soongorica, Rhinopetalum karelinii, Salsola laricifolia, S. rigida, Schismus arabicus, Spirorhynchus sabulosus, Stipa caucasica, S. glareosa, S. orientalis, Tamarix laxa, Trigonella arcuata, Tulipa buhseana.*

IIA (8) Depressio Balchasch-Alakul : *Artemisia kaschgarica, A. terraealbae, Carex pachystilis, Colpodium humile, Eremostachys molucelloides, Lepidium perfoliatum, Poa bulbosa, Tauscheria lasiocarpa, Tulipa kopakovskiana, Veronica campylopoda.*

III Provincia Tibetica

Agropyron thoroldianum, Artemisia wellbyi, Avena subspicata, Chamaerhodos subulosa, Cleistogenes thoroldii, Elymus junceus, E. lanuginosus, Ephedra distachya, E. gerardiana, Eurotia ceratoides, Festuca valesiaca, Kochia prostrata, Microula tibeticus, Oxytropis aciphylla, Muricaria prostrata, Poa alpina, Ptiolagrostis concinna, P.purpurea, Reaumuria trigyna, Rheum spiciforme, Tanacetum gracile, T. tibeticum, T. xylorrhizum, Thermopsis alpina, Thylacospermum rupifragum, Triglochin palustre, species of *Androsace, Carex,*

Cobresia, Corydalis, Cremanthodium, Eriophorum, Gentiana, Heleocharis, Helictotrichon, Ligularia, Meconopsis, Pedicularis, Pegaeophyton, Pleurogyne, Primula, Rhododiola, Rhododendron, Saussurea, Saxifraga.

III A : Tsinghai: (1 Nanschan; 2 Amdo) : *Acanthopanax giraldii, Acer davidii, A. kansuensis, A. philadelphus, A. tetramerum, Agropyron cristatum, A. strigosum, Arenaria formosa, Artemisia caespitosa, A. desertorum, A. frigida, A. macrocephala, Asterothamnus centraliasiaticus, Betula alba-sinensis, B. japonica, Caragana jubata, C. tanguita, Corylus mandschurica, Daphne giraldii, Dasiphora fruticosa, Festuca rubra, F. sulcata, Helictotrichum tibeticum, Hippophae rhamnoides, Hydrangea bretshneideri, Juniperus satuaria, J. tibetica , Kalidium gracilis, Littledalia racemosa, Nannoglottis carpesioides, Picea asperata, P. purpurea, Paraquilegia grandiflora, Pinus tabulaeformis, Poa bulbosa, Pomatosace filicula, Populus cathyana, P. tremula* var *davidiana, Przewalskia tangutica, Reaumuria soongorica, Rheum spiciforme, Rhodiola quadrifida, Rhododemdron anthopogonoides, R. agglutinatum, Saussurea medusa, S. sorocephala, S. stella, Salsola passerina, Sorbaria arborea, Stipa bungeana, S. capillata, S. sibirica, Sibiraea laevigata, Tanacetum achilleoides, T. fruticolosum, Thermopsis lanceolata, Thylacospermum rupifragum, Trollius pumila, Ulmus pumila, Xanthopappus subacaulis.*

III B: Tibet (1 Czantan, 2 Wentzan, 3 Tibet australis): *Acantholimon diapensioides, Androsace tapete, Arenaria musciforme, Artemisia salsoloides, A. welbyi, Astrgalus arnoldii, A. malcomii, A. tibeticus, Buddleia lindleyana, B. tibetica, Caragana jubata, C. tibetica, Carex moorcroftii, Ceratostigma griffithii, Cobresia bellardii, C. roylena, Corylus heterophylla, Cotoneaster acutifolia, Dasiphora fruticosa, Elaegnus pungens, Ephedra distachya, E. gerardiana, Eurotia ceratoides, Heleocharis* sps, *Hippophae rhamnoides, Imperata cylindrica, Juncus thomsonii, Juniperus convallium, J. semiglobosa, J. squamata, J. tibetica, J. prezewalskii, Lonicera spinosa, Meconopsis integrifolia, Myricaria prostrata, Oxytropis aciphylla, Pedicularis rupicola, Phlomis rotata, Polygonum sphaerostachyum, P. viviparum, Picea pseudosabina, Populus alba, P. cathayana, Ptilagrostis concinna, P. purpurea, Prunus ansu, Ranunculus tricuspis, Raeumuria trigyna, Rhododendron fastigiatum, Salix biondiana, S. variegata, Saussurea tridactyla, S. welbyi, Sophora moorcroftiana, S. viciifolia, Stellaria decumbens, S. chamaejsme, Stipa inebrians, Tanacetum tibeticum, Tarpinia arguta, Thermopsis alpina, Thylacospermum rubifragum, Triglochin palustre, Ulmus pumila,* species of *Allium, Atriplex, Cirsium, Eriophorum, Gentiana, Heleocharis, Fritillaria, Primula, Rheum, Saussurea, Salsola, Suaeda, Berberis.*

III C: Pamir : *Acantholimon diapensioides, S. hedenii, A. pamiricum, Arabis kokanica, Artemisia kuschakewiczii, A. leucotricha, A. rhodantha, A. skorniakowii, Astragalus altischuri, A. kuschakewichzii, A. ophiocarpus, Betula microphylla (?) , Braya pamirica, Caragana jubata, Carex melanantha, C. orbicularis, C. pseudofoetida, Christocolea crassifolia, C. pamirica, Cobresia capilliformis, C. pamiroalaica, C. stenocarpa, Comarum salesovianum, Cotoneaster uniflora, Cousinia kokanica, C. rava, Dasiphora dryanthoides, Didymophysa fedczenkoana, Dracocephalum heterophyllum, D. paulsenii, Eurotia ceratoides, E. prostrata, Festuca coelestis, F. sulcata, Gypsophila capituliflorus, Hackelia testimundi, Halimodendron halodendron, Inularhizocephala, Juncus thomsonii, Jurinea depressa, Lysimachia dubia, Lycium rathenicum, Macrotomia euchroma, Nepeta daenensis, N. longibracteta, Oxytropis bella, O. chiliophylla, O. immersa, O. poncinsii, O. tianschanica, Pedicularis cheilanthifolia, Poa glauciculmis, P. litwinowiana, Plantago gentianoides, Ptilagrostis subsessiliflora, Rhamnus minuta, Ribes meyeri, R. janczewskii, R. villosum, Rosularia paniculata, Salix fedtschenkoi, S. pycnostachya, Saponaria griffithii, Sibbaldianthe tetrandra, Smelovskia calycina, Stipa glareosa, S. orientalis, Tanacetum gracile, T. pamiricum, T. tibeticum, T. xylorrhizum, Thylacospermum rupifragum, Tragopogon pamiricum.*

The Mediterranean Region

The Ancient Mediterranean Region (AMR) and the Pamir-Alai in prehistoric period occupied the Tethys Sea in the zone of "contact of holarctis and palaeotropis types of vegetation in the geologic past". The whole of the AMR is considered to be heterogenous by Lavrenko and Ovczinnokov (1971 cf Ovczinnikov et al, 1973) who divided it into three "stripes" or under-zones:

1) The central stripe (Pamir-Alai): dominant deciduous broad leaved forest (relic) and a slight admixture of evergreen species. A number of genera are represented by a single species, otherwise abundant in palaeotropics.

2) Complex relief combined with a significant meridional and longitudinal space of territory: arid, semi-arid and humid micro regions (prec 50-100 mm to 1300-1500 m/yr) and complex florocoenogenetic history of the Pamiro-Alai, flora rich (4500- 5000 species) complex phytocoenotic structure from broad leaved forests and meadows of boreal type to deserts and "djagals".

3) Tadjikistan and Turkestan, many formations belong to dominant ancient Mediterranean flora and some vegetation types typical for it on the one hand and to florocoenotypes of boreal flora on the other.

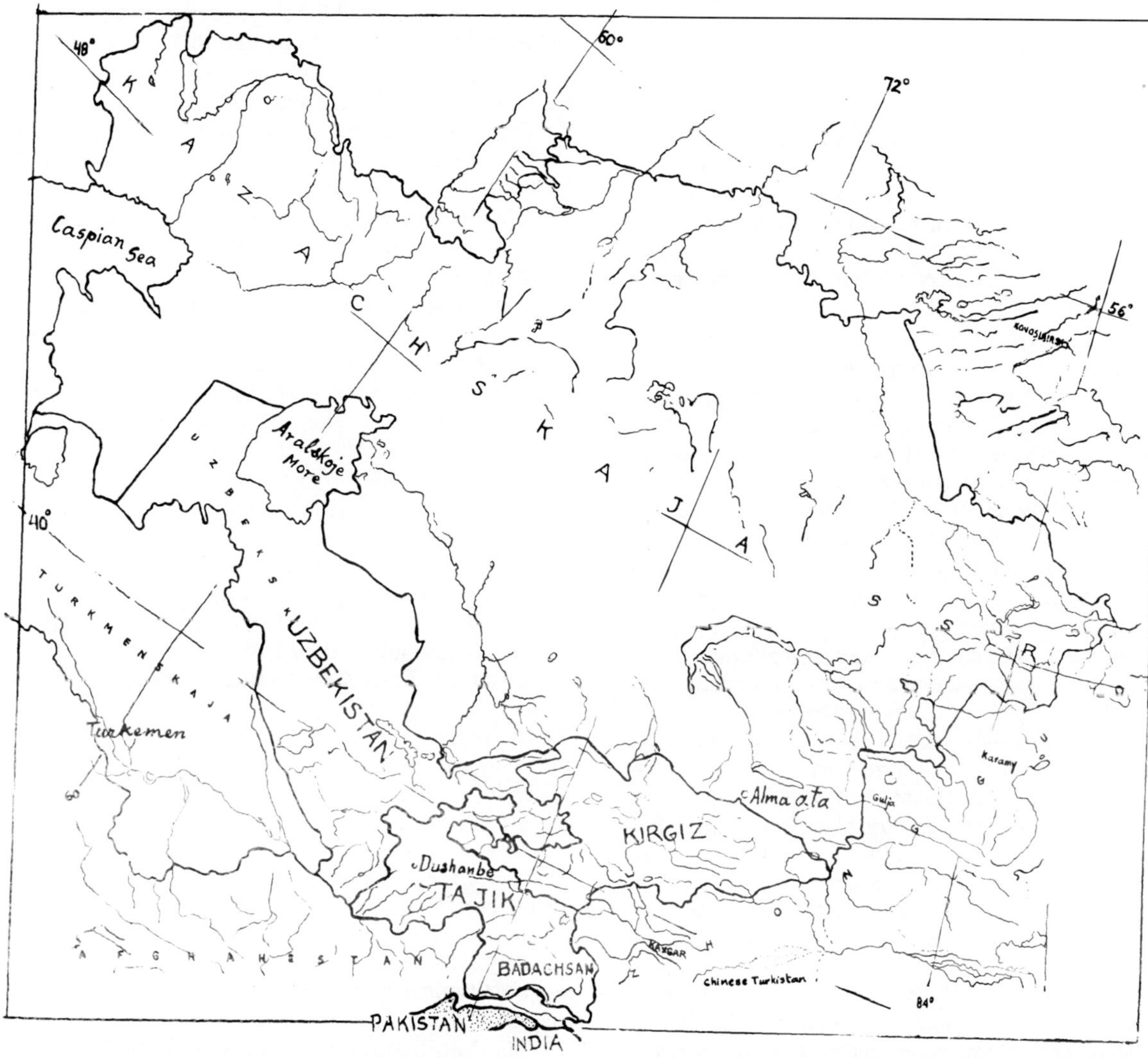

Fig. 13 : The Central Asian republics of the USSR, 1995.

However AMR is connected with the "most grand Neogene- Pleistocene species formation in the vast mountain and plain areas from Iberian Peninsula including westren Himalaya and Turkestan" (Papov, 1950). During the Middle Tertiary the entire Himalaya along with Tibet also occupied the Tethys Sea (Kachroo and Dhar, in lit). The Tethys zone of Kashmir is divisible into two sub-units: a) the valley of Kashmir and b) Karakoram including Ladakh (Pande and Gupta, 1972). The Tethys Sea more or less disappeared in Late Eocene.

Pamiro-Alai

Pamir Alai is a vast system of mountain ranges situated near frontiers of the USSR with Afghanistan and India. It includes the territories of Tadjikstan, south Kirgiz, southwest Uzbekistan and part of Turkemenia (Fig 13). The mountain ranges are Alai- Turkestan, Zeravashan-Gissar, Darvas and Pamir.

In phytogeographic sense the Pamir Alai belongs to Pamir Alai Province of Turkestan sub-region which is a part of a vast phytogeographical region, the AMR, the latter extends from the Hindukush mts and west Himalaya in the east to the Atlas mts and Sierra-Nevada mts in the west (Ovczinnikov, 1970, 1971). "This province borders upon the Central Asian Desert Province, including the East Pamirs Highlands and is part of another phytogeographical region-the Boreal Region".

"The southern part (the southern Pamir-Alai) is similar to the southern subtropical zone of the AMR; this zone is characterised by a very adverse arid climate and high summer temperatures. In the southern P-A light forests with sparse tree stand represented by *Pistacia vera, Ceras griffithii,Amygdalus bucharicus* grasses, *Poa bulbosa, Carex pachystylus, Festuca foetidissima,* the central P-A has warm temperate climate, mesophilous broad leaved forests include *Juglans* regia, Acer turkestanicum, Platanus orientalis and tall forbs and grasses; the northern Pamir Alai is of the Euro-Siberian and Pre- Boreal zone of the AMR, here conifers dominate, specifically Juniperus and abound in alpine bunch grass (steppe), *Festuca sulcata, Poa reflexa,* and along river banks *Betula, Populus, Salix, Hippophae rhamnoides*.

Pamir Alai is rich in flora (4500-5000 species) with a "complex phytocoenotic structure from broad leaved forests and meadows of a boreal type to deserts and djangals (Ovczinnokov et al, 1973). However in Tadjikstan and Turkestan the AMR flora dominate with representation of "flora coenotypes of boreal flora". Among the vertical profile has occurred "cryophylization" of new plant species and forms such as *Poa bulbosa* L., *P. bactriana* Roshev, *P. zaprijagajevii* Ovcz and mountainous species of *Silene, Ranunculus, Prangos,* etc. The local and regional topological conditions result in the following types of vegetation:

(a) mesophytes and hygrophytes: *Poa alpina* L., *Carex microglochin* Wahl, *C. pauciflora* Light F, *Triglochin palustris* L., *Trichophorum lumilum* (Vahl) Schinz et Thell, *Ranunculus affinis* R Br, *Cobresia* sps (those occur in snow steppes).

(b) Xerophytes: *Festuca sulcata* Hack, *Stipa kirghisorum* P Smirm (true caespitose-gramineus steppes); and

(c) deserts: *Ceratoides papposa* Botsch et Ikonn, *Caragana jubata* (Full) Poizm, *Ptilagrostis subsessiliflora* (Rupr) Roshev. in the Tien Shan which is farther to the north in the Nearferganian part of Pamir-Alai the "Vegetation typical of Holarctis is increasing in the high mountains" (Ovczinnikov et al, 1973).

Beside the floristic diversity in Pamir Alai occur a number of endemic species and genera such as *Ostrowskia , Cephalopodus, Chaetolimon, Spyrostegia, Kubitangis, Korshinskya, Cephalorrhizum, Thalaspidium,* etc; and a number of species related to the whole of the Mediterranean area or the parts out of Turkestan territory, such as *Rhus coriaria* L., *Juglans regia,* L. *Punica granatum* L. *Ficus carica* L. *Vitex angus-castus* L. *Paliurus spina-christi* Mill, *Prunus divaricata* Ldb, *Elyatrigia trichophora* (Link) Neski, *Hordeum spontaneum* C Koch, *H. bulbosum* L and abundant ephemeres. To this may be added a host of hemicryptophytes and *Juniperus polycarpus*. They further concluded that botanico-geographic regions of Pamiro-Alai may be a "direct only more xerophylized continuation of the Mediterranean, the most typical characters of which are seen in the eastern Mediterranean part".

In Pamir-Alai Ovczinnikov et al (1973) demarcated a "united botannical and geographical

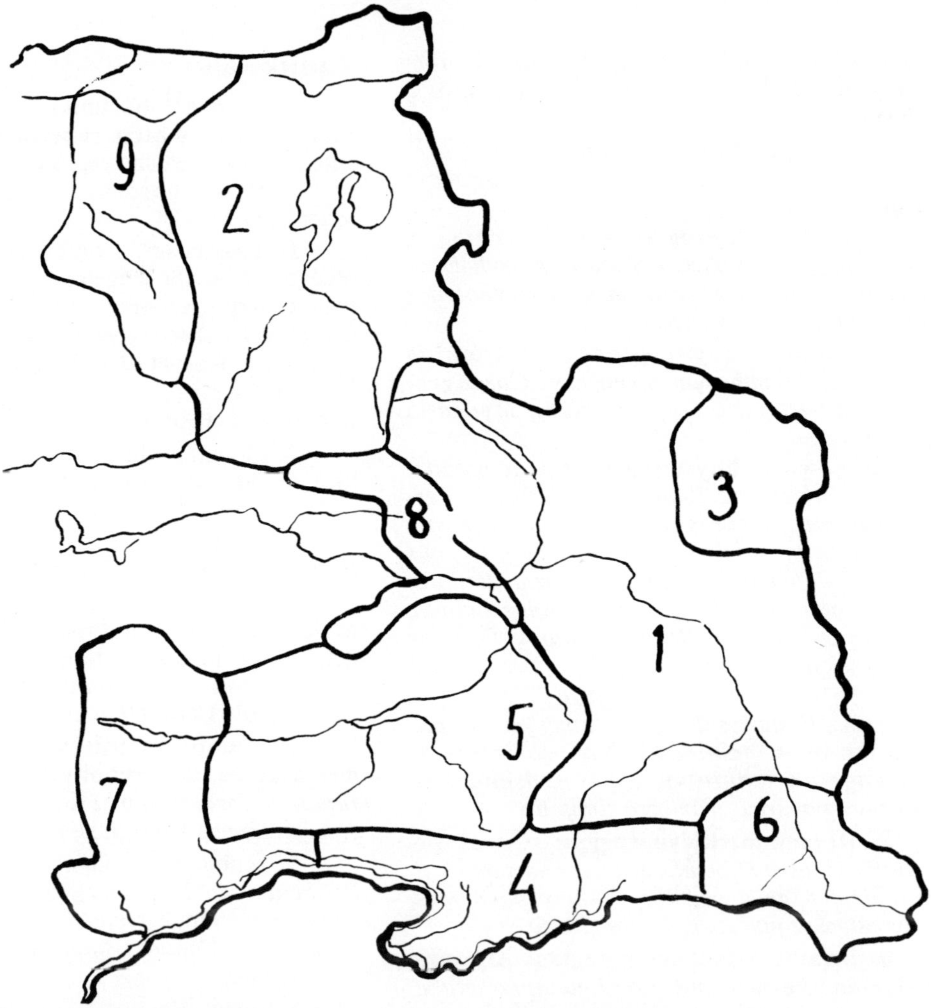

Fig. 14 : The botanical provinces of Pamir-Alai (for explanation see text, cf Ikonnikov, 1963).

district"composed of south Tadjikstan and frontier areas of Afghanistan and Badhyz in Turkeminstan: the west Pamir may also be a district" of the AMR, and east Pamir a part of Central Asian province belonging to the Boreal district. To the Hindukush and Himalaya they relate cissar-Darvas botanico-geographical district "with prevalence of foliage forests, herb steppes, original tragacanth formation and cryptophytons".

On the basis of vegetation Pamiro-Alai has been divided into nine botanical provinces (Fig.14). The representative flora of these is summarised here.

1.Central region: *Eurotia ceratoides, Artemisia rhodantha, Acantholimon diapensoides, Poa dispausa, P. rangkuliensis, Puccinellia akbaitalensis, Torularia brachycarpa, Corispermum gelidum, Salicornia europaea, Salsola paulsenii, Corispermum* (6 sps), *Atriplex, Chenopodium, Sueda, Salsola, Salicornia.*

2. Kara-Kulskiey region: *Christocolea crassifolia, Ch. pamirica, Leymus dasystachys, Oxytropis immersa, Potentilla pamiroalaica, Salix pycnostachya, Astragalus djilgensis, Braya brachycarpa, Tragopogon pamiricus, Anemone gortschakovii, Chorospora bungeana, Saussurea* *leucophylla, Christocolea, Braya, Draba, Torularia, Dilophia, Parrya, Chorispora, Didymophysea, Smelovskia.*

3.Shad-putshkiey region: *Iris loczyi,Ptilagrostis subsessiliflora, Sibbaldia tetrandra, Oxytropis heddinii.*

4. Eor-Kulskiey region: *Artemisia viridis, A leucotricha, Dracocephalum paulsenii, Acantholimon diapensoides, Androsace maxima, Acantholimon alexeenkoanum, Salix caesia.*

5.Alichurskiey region: *Artemisia leucotricha, Xylanthemum pamiricum, Carex griffithii, Poa bucharica, Lonicera olgae, Myricaria squamosa.*

6.Iugo-vostochnyey region: *Carex pseudofoetida, C. orbicularis, Kobresia capilliformis, Festuca sulcata, Artemisia rhodantha, Agrostis pamirica, Erithrotrichum pseudostrictum, Thymus diminutus, Potamogeton natans, Schulzia albiflora, Acantholimon tianschanicum, Rhamnus minuta, Cotoneaster uniflora, Rosa webbiana, Berberis kashchgarica.*

7.Bulun-Kulsiey region: *Artemisia leucotricha, Festuca sulcata, Piptatherum laterale, (Acantholimon pamiricum), A. varivtzevae, Potamogeton tubulatus, Salix fedschankoi, Semenovia pamirica, Winklera silaifolia.*

8. Djamantalskieyu region: *Carex orbicularis, C. pseudofoetida, Salix pycnostachya, S. schuanica, Ajanis gracillis, Rumex pamiricus, Psatyrostachys juncea, Ephedra glauca.*

9. Belaind - Kukskiey region: *Artemisia rhodantha, Festuca sulcata, Leucopoa olgae, Alopecurus hymalicus, Festuca tianschanica, Anaphalis virgata, Ribes villosum, R. jantczevskii.*

In the dendroflora, 267 species are included and the dominant genera are the following (* no of species in each):

Astraphaxis, 8*	*Juniperus*, 4
Berberis, 21	*Lonicera*, 18
Calligonum, 14	*Populus*, 9
Caragana, 6	*Rosa*, 17
Crataegus, 11	*Salix*, 13
Ephedra, 8	*Tamarix*, 9

Tadjikstan

Tadjikstan[11] occupies central part of the Pamiro-Alai which represents 90 % of its geographical area and represents northeastern part of the AMR. In the south it abuts on the Hindukush mountains.

In Tadjikstan 20 florocoenotypes have been recognised as the vegetation has not " suffered post-tertiary glaciation and lost its florocoenotic complexes". These types are placed in three groups by Ovczinnikov et al (1973) and are reproduced verbatim here.

I. Tree and shrub vegetation

1. Mountain thermophyllous and mesophyllous deciduous broadleaved forests. derivat of arctotertiary or turgai flora of the past:

Juglans regia L.,*Acer turkestanicum* Pax, *Platanus orientalis* L. *Exochorda alberti* Rgl.

2*. Mountain mesophyllous forests, small leaved oligothermic and microthermic birch forests, disjunctive in distribution: *Betula* div species, *Populus tadschileistanica* Kom, *Fraxinus potamophila* Herd, *Hippophae rhamnoides* L.

3+. Juniperus forests: subtypes: 1) thermophyllous, 2) cryophyllous, 3) *Juniperus: Juniperus seravschanica* Kom, J. *semiglobosa* Rgl, *J. turkestanica*, Kom, *J. sibirica* Burgsd.

4.+ Shibljak-hemixerophyllous sparse forests: *Pistacia vera* L., Amygdalus div, sps. *Celtis caucasica* Willd, *Zizyphus jujuba* Mill, *Cartis griffithii* Boiss, *Paliurus spinachristi* Mill, *Acer* div species, etc.,

5. Tugai: *Populus pruinosa* Schrenk, *Elaegnus orientalis* L.

6. Mountain taiga: Djangal: *Haloxyon persicum* Bge ex Boiss, *Salsola ruchteri* Kor, *Ephedra strobilacea* Bge, *Caligonum* div sps.

II. Shrublets, semishrublets and high herb vegetation

7. Deserts

subtypes a) deserts of the AMR type: Artemisia div sps, Saslola rigida Pall, S. arbuscula Pall.

11. elevation 300-6000(7000) m.a.s.l., climatic regime: sharp continuality of mediteranean type; heterorythmic precipitation, mild winters, hot summers.

b) deserts of the 'cold 'boreal Central Asian type: *Ceratoidea papposa* Botsch et Ikonn, *Artemisia rhodhantha* Rupr, *Ajanica tibetica* (HK. f & T) Thaval.

8. Halophyton: *Halostachys belangeriana* (Moq) Botsch, *Halocnemum strobilaceum* (Pall) MB, *Sieditzia zosmazinus* (Ehrh) Bge.

9. Tragacanth or tragacantniki:

a) thorn cushion plant formation: *Onobrychis echidna* Lipsky, *Astragalus* sect Aegacantha, *Acantholimon* div sps.

b) thorn herbaceous plant formation: *Cousinia* div sps.

10. Thymus communities, analogous to mediterranean tomillars:

Perovskia scrophelariifolia Bge, *Thymus seravschanicus* Klok, *Zizyphus pariroalaica* Juz, *Artemisia persica* Boiss.*Hypericum scabrum* L.

11. Tsbalniki-plant communities:

Alhagi div sps, *Lagonichium farctatum* (Backset Soland) Bobr, *Glycyrrhiza* div sps.

12. Pseudohammada vegetation: *Cephalorrhizum oopodum* M Pop et Korov, *Chaetolimon* div sps, *Cleome neoana* Boiss, *Capparis rosanoviana* B Fedtsch. *Astragalus bucharicus* Rgl, etc (formations on mosaic gypsum soils).

III. Herbaceous vegetation

13. Semisavannas

a) low herbs: *Carex pachystylus* J Gary, *Poa bulbosa.*

b) high grasses: *Elytrigia trichophora* (Link) Nevski.

c) high herbs: *Prangos pabularia* Lindl, *Ferula* div sps, *Inula grandis* Schrenk ex Fisch et Mey, *Crambe kotschyana* Boiss, *Ligusticum discolor* Ldb

d) ephemeretum of the species: *Teaniatherum, Aegilops, Bromus, Anisantha.*

14. Steppe

a) caespitose-gramineous : *Festuca sulcata* Hack, *Stipa* div sps, *Poa reflexa* Ovcz, *Leucopoa karatavica* (Bge) V Krecz et Bobr, *Halictotrichon hissaricum* (Roshev) Henr, etc.

b) herba diversa: *Artemisia dracuncula* L. *Adonis turkestanica* (Karsh) Adolf, *Nepeta podostachys* Bth.

15⁺. Savannoides: *Erianthus revennae* (L.) PB, *Imperata cylindrica* (L.)PB, *Saccharum spontaneum* L.

16*. Meadows: *Poa pratensis* L. *Alopecurus seravschanicum* Ovcz, *Ligularia thomsonii*(Cl.) Pojark, *Polygonum coriarium* Grig.

17. Saz: *Carex* div sps, *Trichophorum pumilum* (Vahl) Schinz et Thell, *Cobresia* div sps.

18. Herbaceous swamps: *Phragmites communis* Trin.

19⁺. Cryophyton: *Puccinella subspicta* V Krucz, *Oxytropis* div sps. *Androsace caduca* Ovcz, *A. akbaitalensis* Berg, *Ermania flabellata* (Rgl) OES, *Corydalis fedtschenkoana* Rgl. *Ranunculus* div sps. *Potentilla* div sps.

20. Petrophyton: Some species of *Dionysia, Scutellaria, Nepeta, etc.*

They further grouped the types* 1,2,16 and 19 as mesophylic; and types⁺ 3,4,14, (15,19) as hemixerophytes. Type 19 was also considered as a "special physiological xerophyte".

Turkestan is mainly mountainous and semisavannas are typical of its southern part (with some breaks up to the western Mediterranean). Its montane character retains the heterochroneous elements of the past, as well as vegetation represented by: 1) ephemerization, 2) tragacantization, 3) futicization, 4) futiculization (*Artemisia, Salsola, Polygonum*), 5) oreophylization (cryophyllization, petrophyllization). " The distribution of the contemporary vegetation in Turkestan is in accordance with the laws of vertical zonality".

The Botanical regions of Kazakhistan are given in Fig. 15.

Western and Central Asian region

Takhtajan (1969) following Good in classification (cf Virjee 1983) recognised a " Holarctic Kingdom" dividing it into three subkingdoms, namely (a) the Boreal, (b) the Tethyan or Ancient Mediterranean and (c) the Madrean or Sonaram. He limited the Tethyan subkingdom to the old world only but included in it the Macronesian and Irano-Turanian regions. In 1974 in his Irano- Turanian

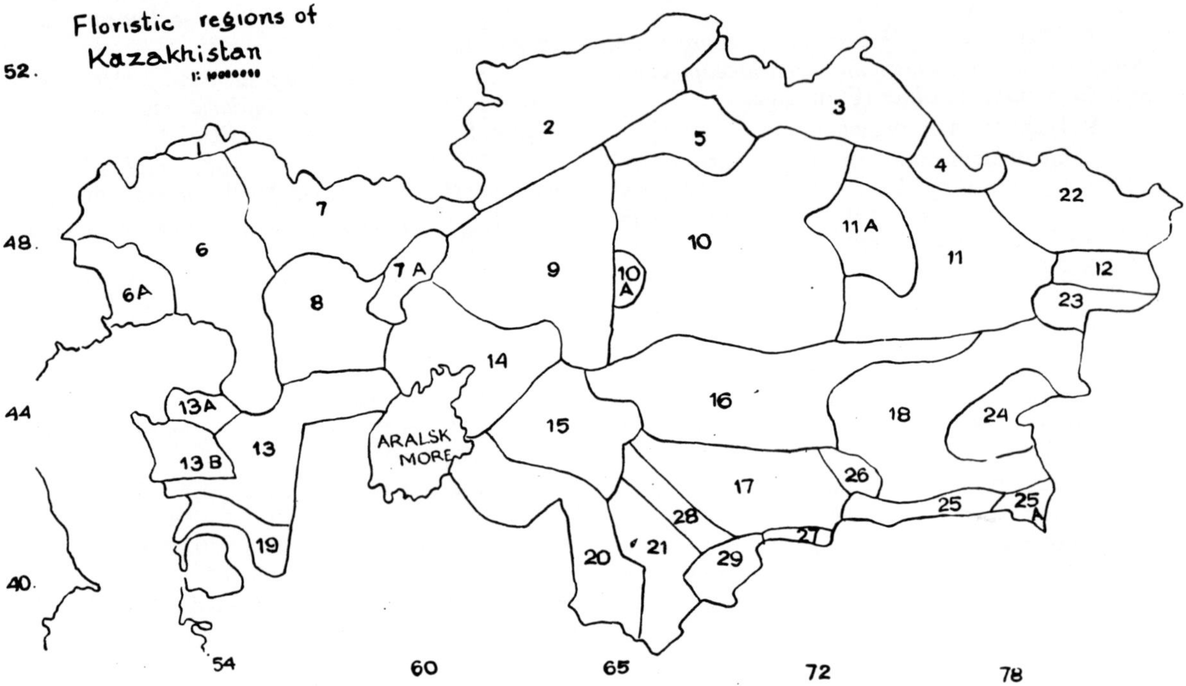

Fig. 15 : The floristic regions of Kazakhstan (cf Flora Kazakh, Vol 1).

1 spurs of a common syrt, 2. tobolsk-isimsky r, 3. irtyshsky r, 4. semipalatinsky borosoi r, 5. kokchetansky 6. prikspiishy r. 6a bukeyssky r, 7. aktyubinsky r, 7a mugadzhary, 8. embensky 9. turgaisky r, 10. west small-(cone-shaped) hill, 10a ulutau, 11. east small (xone-shaped) hill, 11a karkaralinsky r, 12. zaisansky, 13. northern ust-urt, 13a buzachi, 13b margyschlak 14. prairalsky, 15. kzyl ordinsky 16. betpakdalinsky,17.muyun-kumsky, 18. b;ljasj-alakulsky, 19. south ust-urt. 20. kzyl-kumsky, 21. turkistansky, 22. altai, 23. tarbagatai 24. dzhungarsky ala-tau, 25a zailiisky kungei alatau, 25a ketmenterskei alatau, 26, chu- iliisky mts, 27. kirghiz alatau, 28. karatau, 29. western tyan- shan.

region, he recognised a " western asiatic subregion". This includes the entire western part of Irano-Turanian region up to the Prebalkhas area, central Tien Shan and western Himalaya. The 8th province here is Western Himalaya Province. Here he included " Nuristan, Kafristan, valleys of the rivers Kabul, Kurram, Kunar, Swat and Gilgit, Wazirstan, and southern slopes and off-spurs of western Himalaya, westwards approximately from 83° E long. to an average of 1000 m.a.s.l. Kashmir, Simla, Mussurie and Nainital belong to this province!

A considerable part of this province he stated is to be characterised by monsoon climate: *Parrotiopsis, Myrtus communis, Cedrus deodara, Abies spectabilis, Abies pindrow, Picea smithiana, Pinus wallichiana, P. gerardiana* (giving areas under each). *Cupressus torulosa* he stated to be occurring from Kashmir to central Nepal. It does not occur in Kashmir.

"During Pleistocene, the Central Asian mountains and the Himalaya were linked with the extended arctic girdle. Migration of species could easily occur . . stock of species during Teriary . . . as the ancestors of all the xerophytic diverse taxa and schizoendemics . . similar Teriary ancestors must also exist for the alpine flora . . and there are floristic links across the Eurasian continent from the Pyrennes, the Alps: the Carpathians, the Caucasus to the central asiatic mountains, Altai to the Himalayas" (Breckle, 1974).

Dobremez (1972) commented that "west Himalaya, including Kashmir and parts of Hindukush with dry climate of the Mediterranean type excluding most of the monsoon influence are different from the desert climate of the mountains of Afghanistan".

Raven (1971) stated that (a) Mediterranean type areas of northern hemisphere have obtained their flora from a mixture of temperate and tropical ancestors and (b) in the southern hemisphere the flora has evolved completely from the tropics; and that similarity between the Mediterranean flora of South Africa and southwest Australia may be attributed to the differentiation of the floras from a common tropical ancestor.

The Mediterranean territory of the Middle East does not seem to have served as an independant centre of speciation, though 50% of the endemics occur in this region. Zohary (1971) has suggested that most of the herbaceous flora is derived from the Irano-Turanian stock. It may have supplied many derivatives to the Saharo-Arabian.

Meusel and Schubert (1971) have concluded that (a) the woody flora of the west Himalaya is remarkably related to the Mediterranean region and many a taxa or closely related trees and shrubs from this region grow in the inner valleys of the west Himalaya which are most strongly influenced by monsoon winds but are open to the west wind drift in winter. (b) However, such tree and shrub elements in the Himalaya are not outposts of the Mediterranean centres but rather fragments of an old semihumid, semiarid, Eurasian flora derived from humid ancestors. In other words west Himalaya contains not only outposts of Mediterranean Turkestan taxa but also fragments of an old Mediterranean (Tethys) flora.

Zohary (1971) discussed the floristics of Middle East and divided it into five main territories: the Euxine-Hyrcanian,Irano-Turanian, Saharo-Arabian and Sudanian. The flora of Iran according to him (1963) bridges the gap between the four major phytogeographical regions: the Irano-Turanian, the Euro-Siberian, the Saharo-Arabian and the Sudanian. It is also one of the largest speciation centres of the Holarctic desert flora; being rich in floral wealth (ca 7000 plant species) and includes a high number of Irano-Turanian endemics; besides many species with bi-or pleuriregional distribution. Incidetally Wandelbo (1971) included in the Iranian regime plants of Iran, Afghanistan, adjacent mountains of Iraq, Turkeminstan and even western Pakistan! He also noted that Mediterranean influence was vivid in the flora of the region so demarcated and that there was justification for creation of subregions for Iran and Afghanistan and that Afghan flora shows a closer affinity with that of Central Asian flora. The Koper Dagh mountains include both the Central Asian and the Mediterranean elements.

GENERA WITH SINGLE SPECIES

Kazakhstan

SPECIES NOT COMMON WITH ANY ADJACENT REGION*

Acinos graveolens (MB) Link
Achyrophorus maculatus (L) Scop
Acorellus pannonicus (Jacq) Palla
Aesulus hippocastatum L.
Aflatunia ulmifolia (Franch) Vass
Ajuga pseudochia Schost
Alliaria officinalis Andrz
Alsine segetalis L
Amythysiea coerulea L
Anagallidium dischotomum (L) Griseb
Ancathia igniaria (Spreng) DC
Andrachne rotundifolia CAM
Anisum vulgare Gaertn
Antennaria dioica (L) Gaertn
Antirrhinum majus L
Arabis tianschanica N Pavl
Arctogeron gramineum (L) DC
Arctous alpina (L) Niedenzu
Aristolochia clematis L
Ascarium europaeum L
Asterothamnus fruticosus (Winkl) Novopokr
Beta vulgaris L
Bienertia cycloptera Bge
Braya siliquosa Bge
Cacalia hastata L
Coccinia dubia Bge
Calamogrostis karataviensis P Smirn
Calendula officinalis L
Calluna vulgaris (L) Hill
Callistephus cinnensis (Juss) Rothm
Cancriniella krascheninnikovii (Rubtz) Tzvel
Carlina biebersteinii Bernh
Carpoceras ceratocarpum (Pall) N Buch
Castilleja pallida (L) Kunth
Cenolophium fischeri (Spring) Koch
Cerintha minor L
Chaerophyllum prescottii DC
Chamaegeron oligocephalus Schrenk
Chamaesphacos ilicifolius Schrenk
Chaiturus marrubiastrum (L) Spenn
Chartolepis intermedia Boiss
Chelidonium majus L
Chimaphila umbellata (L) Nutt
Cinna larataviensis N Pavel
Claytonia doanneana Roem et Scult
Coeloglossum viride (L) Hartm
Conioselinium vaginatum (Spring) Thell
Convallaria majalis L
Cryptocodon monocephalus (Trautv) Fed
Crupina vulgaris Cass
Cymbaria dahurica L
Cynara scolymus L
Dryas oxydonta Juz
Elacopleurum monococcum Kor
Elscholzia patrinii (Lap) Garcke
Empatrum subholarcticum V Vassil
Erianthus purpurescens Anderss
Eriosyanaphe lignifolia (Fisch) DC
Erythronium sibiricum (F et M) Krvl
Eupatrium cannabinum L
Evax flaginoides K et K
Falcaria sioides (Wib) Aschers
Farsetia spathulata K et K
Ficaria verna Huds
Gymanthus gamocarpus (Moq) Bge
Globularia aphyllanthes Cr
Glycine hispida (Moench) ? Maxim
Gymnadenia conopsea (L) R Br
Gymnanthus gamocarpus
Halotis pilosa (Moq) Iljin
Herminium monorchis (L) R Br
Hepatica falconeri (Thoms) Juz
Heteracia szovitsii F et M
Hololachne soongoria (Pall) Ehrh
Holoschoenus vulgaris Link
Holosciadium nodiflorum (L) Koch
Hypopitys monotropa Crantz
Iberis amara L

---

* adjoining regions : Kazakhstan, Uzbekistan, Tadjikstan, Kirgiz.

Ikonnilovia kaulfmanniana (Rgl) Lincz
Ipomoea purpurea (L) Ruth
Iris falcifolia Bge
Kaulfmannia semenovii (Herd) Rgl
Knautia arvensis (L) K Coult
Koroviniatenuisecta (Rgl et Schmath) Nevski et Vved
Kraschenin nikowia rupestris Turcz
Krasnovia longiloba (Ket K) MP
Lagosariopsis papovii (Krasch) Kirp
Leontodon autumnalis L
Leptopyrum fumarioides (L) Rchb
Lilium martagon L
Lindernia pyxidaria All
Linnaea borealis L
Liparis loeselii (L) Rich
Lysimachia vulgaris L
Macropodium nivale (Pall) R Br
Majanthemum bifolium (L) Schimidt
Malachrum aquaticum (L) Fries
Mausolea eriocarpa (Bge) Pojark
Meristotropis erythrocarpa Vass
Metastachya sagittata (Rgl) Knorr
Middendorfia borysthenica (MB) Trautv
Mollugo cerviana (L) Ser
Microstylis monophyllos (L) Lindl
Myosurus minimus L
Nasturtium fontanum (Lem) Aschers
Naumbergia thyrsiflora (L) Rchb
Neottianthe cucucculata (L) Schlechter
Nicandra physaloides (L) Gaertn
Niedzvedzkia semietschenskia B Fedt
Nymphoides peltatum (Gmel) OK
Odontites serotina (Lam) Dun
Oenanthe aquatica (L) Poir
Olgae pectinata IIjin
Onagra biennis (L) Scop
Ornithogalum fisderianum H Krasch
Crthantha lutea (L) Kern
Osmorhiza cristata (Thunb) Mak et Yabc
Ostericum palustre Bess
Paramicrorhynchus procumbens (Roxb) Ki
Paris quadrifolia L
Parrya longicarpa Krassen
Pastinacopsis glacialis Golosk
Persica vulgaris Mill
Perovskia abratanoides Kar
Petasites spurius (Retz) Rchb
Phaecasium pulchrum (L) Rehb
Phalacrachena calva (Ldb) Iljin
Physandra halimocnmis Botsch
Pilostemon filifolia (Winkl) Iljin
Pilopleura kosopoljanskii Schischk
Platanthera bifolia (L) Rich
Pleurospermum uralense Hoffm
Prionotrichon erysimoides (K et K) Botsch et Vved
Ptilotrichum tenuifolium (Steph) CAM
Rhaphidophyton regelii (Bge) Iljin
Rhus hirta (L) Sudw
Rhynchospora alba (L) Vahl
Saponaria officinalis L
Scariola orientalis (Boiss) Pojak
Scleropoa rigescens (Trin) Grossh
Scolochloa fastucacea Link
Schumannia karelinii (Bge) Kor
Scleriotiana pentaceras Kor
Semenovia transilienisis Rgl et Herd
Seselopsis tianschanicum Schischk
Sergia severzovii (Rgl) Fedt
Sibbaldianthe adpressa (Bge) Juz
Sinapsis atavensis L
Soranthus meyerii Ldb
Spergula vulgaris Boeningh
Spiraenthus schrenkiana (F et M) Maxim
Spiranthes amoena (MB) Spreng
Spirodela polyrrhiza (L) Schleid
Sterigmostemum tomentosum MB
Stevenia cheiranthoides DC
Stiziophus balsamita (Lam) Cass
Strogonovia desertorum (Schrenk) Botsch
Stratioites aloides L
Succisa pratensis Moench
Symphoricarpus albus (L) N Blake
Syringa vulgaris L
Thaldiantha dubia Rgl
Thevenotia scabra Boiss
Thuspeinatha persica (Boiss) Briq

Tillaea vailantii Willd
Trachydium tianschanicum H Kor
Tragus racemosus (L) Desf
Trogonotis peduncularis (Trev) Bth
Trientalis europaea L.
Tropolium vulgare Nees
Tunica stricta (Bge) F et M
Ungernia severzovii (Rgl) B Fedt
Vigna sinensis (L) Endl
Viscaria viscosa (Scop) Aschers
Xeranthemum longipapposum F et M
Xiphium kolpakovskia (Rgl) Baker
Zinnia elegans Jacq
Zosima absinthifolia (Vent) Link
Zostera minor (Cavol) Nolte
Zygophyllum hudunense Semiotr

SPECIES COMMON TO KAZAKHSTAN AND UZBEKISTAN

Acantholepis orientalis Less
Acorus calamus L.
Aldrovania vesiculosa L.
Alexandra lehmnni Bge
Ambrosia artemissiforlia L
Athamanta macrophylla (Rgl et Schmath) Kor
Antonina debilis (Bge) Vved
Bongrdia chrysogonum (L) Spachp[1]
Borsczowia aralocaspica Rom
Buchingera axillaris Boiss[1]
Campylopteris carnea (Banks et Soland) Botsch et Vved.
Chalacantha renifolius (Boiss) Boiss
Chamasphacus illicifolius Schrenk
Chloris virgata SW
Cressa cretica L
Cuminum cyminum L
Cylindrocarpa secertzovii (Rgl) Rgl
Dahlia pinnata Cav
Echallium elaterium (L) Rich
Echinaria capitata (L) Desf
Eriocaulon sieboldianum Sieb et Zucc
Eriochloa succincta (Trin) Kunth
Eruca sativa Mill[1]
Ewersmannia subspinosa (Fisch et DC) B Fedtsch
Foenicum vulgare Mill
Frangula alnus Mill
Genista tinctoria L
Gratiola officinalis L
Halopeplis pygmaema B (Pall) Bge ex Ung-Sternb
Heliopsis scabra Dun
Hymenolaena pimpinellifolia Rupr
Iberidella trinervia (DC) Boiss
Lapsana communis L
Listera ovata (L) R Br
Londesia erianthus F et M
Melacocarpus crithmifolius (Retz) CAM
Merendera robusta Bge
Microphysa elongata (Schrenk) Pobid
Ocimum basilicum L
Octoedras lehmannianum Bge[1]
Ojarston monandrium (Pall) Moq
Pseudohandelia umbellifera (Boiss) Tzvel
Pseudohnosyris grimmii (Rgl et Schmath) Novopokr.
Queria hispania L[1]
Rhabdotheca korovinii (MP) Kirp
Russovia sogdiana (Bge) B Fedtsch
Salicornia europea L.
Schiskinia albispina (Bge) Iljin
Spirohynchus sabulosus K et Kp[1]
Stepanicaryum olgae (B Fedtsch) MP
Streptoloma desertorum Bge[1]
Sauchtelenia calyciana (CMA) DC
Tauscheria lasiocarpa F et M[1]
Tetradictylis tenella (Ehr) Litv
Trachyspermum ammi (L) Sprague
Trichochiton inconspicuum Kom
Viburnum opulus L.

ADDITIONAL SPECIES COMMON TO KAZAKHSTAN AND TADJIKSTAN

Abutilon theophrasti Medic
Armeniaca vulgaris Lam
Buschia talerifolia (DC) Ovcz

Chalcanthus renifolius
Coriandrum sativum L
Cressa cretica
Daucus carota L
Descurainia sophia
Echinophora sibthorpiana Guss
Eranthus longistipitata Rgl
Eriochloa succuncta
Eriocaulon sieboldianum Sieb et Zucc
Eruca sativa
Euclidium syriacum (L) R Br
Foenicum vulgare
Glaux hispida (Moench) Maxim
Gymnocladus dioca (L) Koch
Halostachys belangeriana (Moq) Botsch
Hippuris vulgaris L
Juncella serotina (Rottb) CB Cl
Lloydia serotina (R) Rchb
Londesia eriantha
Maclura amantiaca Nutt
Microgynoecium tibeticum HK f
Neottia camtschatea (L) Rchb
Oxyria elatior R Br
Ricinus communis L
Schoenoxiphium hissaricum Pissjauk
Spirorhynchus sabulosus K et K
Sympegma regelii Bge
Tetradictylis tenella (Ehrenb) Litv
Vaccaria pyramidata Medic
Ulugekia tschunganica (B Fedt) Zak

ADDITIONAL SPECIES COMMON TO KAZAKHSTAN AND KIRGIZ

Ageratum houstonianum Mill
Anethum graveolens L
Aphaenopleura cappilifolia (Rgl et Schmath Lipsky)
Bellis perennis L
Borago officinalis L
Capparis spinosa L
Catabrosa aquatica (L) PB
Celtis caucasica Willd
Chaetolimon setiferum (Bge) Lincz
Cichorium intybus L
Cicuta virosa L
Corallorrhiza trifida Chatel
Cynanchium sibiricum Willd
Dictamnus angustifolius G Don
Glechoma hederacea L
Greditschia triacanthos L
Hachelia deflexa (Wahlb) Opiz
Humulus lupulus L
Hypogomphia turkestana Bge
Juncellus serotinus
Koenigia islandica L
Kosopoljanskia turkestanica Bge
Lavatera thuringiaca
Leersia oryzoides (L) SW (+ Kash)
Panderia turkestanica Iljin
Parentucella flavifora (Boiss) Nevski
Pentanema albertoregalia (Wink) Gers
Plagiobasis centauroides Schrenk
Quercus robur L
Scilla puschkinioides Rgl
Tilia cordata Mill
Trachylanthus korolkovii Lipsky

ADDITIONAL SPECIES COMMON TO KAZAKHSTAN, UZBEKISTAN AND KIGIZ

Agrostemma githago. L
Alnus glutinosa (L) Gaertn
Amorpha fruticosa L
Arceuthobium oxycedri (DC) MB
Armoracia rusticana (Lam) Gaertn,Mey et Scherb
Arum korolkovii Rgl
Asperugo procumbens L
Brachyactis ciliata Ldb
Bungea vesiculcifera (Herd) N Pavl
Butomus umbellatus L
Chardinia orientalis (L) Ktze
Cichorium intybus L
Cladium mariscus (L) R Br
Clinopodium integgerinum Boriss
Cnicus benedictus L
Codonopsis clematidea (Schrenk) CB Cl

Corylus avellaria L
Cotinus coggygria Scop
Crytospora falcata K et K
Cymbolaena griffithii (Boiss et Rent) Soljan
Cynomorum songaricum Rupr
Diptychoparpus strictus (Fisch) Trautv
Dodartia orientalis C Koch L
Drepanocaryum severzovii (Rgl) Pojark
Echinacea purpurea (L) Moench
Eremodaucus lehmannii Bge
Galinsoga parviflora Cav
Halocharis hispida (CAM) Bge
Halocnemon strobilaceum (Pall) MB
Handelia trichophylla (Schrenk) Hiemerl
Hemerocallis flava L
Hippophae rhamnoides L
Ixilirion tataricum (Pall) Roem et Schult
Karelinia caspica (Pall) Vass
Kickxia elatina (L) Dum
Korolkovia severzovii Rgl
Lachnophyllum gossypium Bge
Lallemantia royleana Bth
Lepidolipsis turkestanica (Rgl et Schmath) Polia
Leptunis trichoides (J Gay) Schisch
Leptaleum filifolium (Willd) DC
Leucanthemum vulgare Lam
Lycopsis orientale L
Melissa officinalis L
Meniocus liniifolius (Steph) DC
Ostrajskia magnifica Rgl
Petroselinium crispum L
Pecnomon acarna (L) Cass
Pistacea vera L
Psoralia drupacea Bge
Saccharum spontaneum L
Schmalhausiana nidulans (Rgl) Petra
Siegesbeckia orientalis L
Sphaerophysa sabula (Pall) DC
Stephanocaryum olgae (B Fedt) MP
Tussilago farfara L
Ugamia angrenia (Krasch) Tzvl
Vallisneria spiralis L
Velezia rigida L
Zizyphus jujuba Mill (Pall)

Uzbekistan

SPECIES NOT COMMON WITH ADJOINING AREAS AND KASHMIR

Acroptilon repens (L) DC
Ajania kokanica (Krasch) Tzel
Alberta paleacea Rgl et Schmath
Amelopsis vitifolia (Boiss) Planch
Angelica ternata Rgl et Schmath
Anthochlamis polygaloides (F et M) Fenzl
Anthriscus sylvestris (L) Hoffmm
Anura pallidivirens (Kult) T Schern
Aristolochia macrophylla Lam
Asthenatherum forskohleri (Vahl) Nevski
Barbaria arctuata (Opiz ex J et C Presl) Rchb
Bassia hyssopifolium (Pall) O.K
Beckmannia eruciformis (L) Host
Bergenia hissarica Boriss
Bidens tripartita L.
Biflora testiculata (L) Spreng
Buxus sempervirens L.
Caesalpinia gilliesii (Wall ex Hk) Bth
Campsis radicans (L) Seem
Carpinus betulus L.
Cephalorrhizum oopodum MP et Korov.
Chamaenerium angustifolium (Bge) Bge
Chartloma platycarpium (L) Holub
Cornulaca korschinskyi Litv
Cortusa turkestanica Losinsk
Cousiniopsis atractyloides (C winkl) Nenvski
Crotolaria juncea L
Cymbolaena griffithii (Gray) Wagenitz
Cystidospermum cheirolepis (F et M) Prokh
Dionysia hissarica Lipsky
Dipterocome pusilla F et M
Eleusine indica (L) Geartn
Eremopoa oxyglumis (Boiss) Roshev
Eutremia integrifolium (DC) Bge
Ferula nuratavica MP
Fritillaria eduardii Rgl
Fumariola tukestanica Korsch
Galeopsis ladanum L.
Gnaphalium luteoalbum L.

Gomphocarpus fruticosus (L) Ait
Goniolobium severtzovii Herd
Gontocharovia papowi (B. Fedtsch et Gontsch) Boriss
Halostachys caspica (Bieb) CAM
Halotis pilifera (Moq) Botsch
Hesperis silurica L
Heteroderis pusilla (Boiss) Boiss
Hippolytia megacephala (Rupr) Poljak
Hosta plantaginea (Lam) Aschers
Hyacinthus orientalis L
Hyssopus zeravschanicus Pazij
Indigofera tinctoria L
Iris maricoides Rgl
Jasminum fruticans L
Kentranthus ruber (L) DC
Kerria japonica (L) DC
Komoravia anisosperma Korov
Lagonichium farcatum (Banks et Sol) Bobr
Lagotis korolkovii (Rgl et Schnath) Max
Lepichiniella seravschanica (Lipsky) MP
Lespedeza bicolor Turcz
Lindernia procumbens (Krock) Bord
Lipskya insignis (Lipsky) Nevski
Lipskyella annua (C Winkl) Juz
Liriodendron tulipifera L.
Lupinus polyphyllus Lindl
Mausolea eriocarpa (Bge) Poljak
Mariscus humulosus (Bied) Hooper
Mazus japonicus (Thunb) O.K.
Meristotropis triphylla (F et M) F et M
Mesostemma karatavica (Schischk) Vved
Moeringia umbrosa (Bge) Fenzl.
Myriactis wallichii Less
Nitraia sibirica Pall
Nymphaea candida J et C Presl
Omalotheca supina (L) DC
Oenothera biennis L
Origanum tytthanthum Gontsch
Orlaya grandiflora (L) Hoffm
Parnassia lazmannii Pall et Schult
Parrotia persica CAM
Parilla nankinensis (Lour) Decne
Periploca graeca L[1]
Phlomidoschema parviflorum (Bth) Vved
Physochlaena alaica Korotok ex Kovalevsk
Phytolacca americana L
Polanisia dodecandra (L) DC
Polychrysum tadschikorum (S Kudr) Kovalevsk
Polytaxis lehmannii Bge
Psathyrostacys juncea (Fisch) Nevski
Psilurus incurvus (Gouan) Schianz et Thell
Pterocephalus afghanicus (Aitsch et Hamsl) Boiss[1]
Pulsatilla campanella F et Rgl et Til
Pyracantha coccinea (L) Roem
Rhizocephalus orientalis Boiss
Rageretia thea (Osbeck) MC Johmst
Sambucus nigra L
Satureja hortensis L[1]
Seidlitzia rosmarinus Bge
Spathipappus griffithii (CB Cl) Tzvel
Sphaerosciadium denanaense (Schischk) MP et Klujuykov
Spirostegia bucharica (B Fedtsch) Ivan
Spryginia winkleri (Rgl) MP[1]
Stachyopsis oblongata (Schrenk) MP et Vved
Stenbergia fischeriana (Herb) Roem
Syreischtschikovia spinulosa (Franch) Pavl
Syremia monana (Pall) Klok
Thellungiella salsuginea (Pall) OES[1]
Thevenotia torulosa (Desf) OES
Triglochin palustre L
Trogonotis peduncularis (Trev) Bth ex Baket et S Moore
Tropoeleum majus L
Tuaniphytum eranthemum (Bge) Poljak
Vaccaria hispanica (Mill) Kauschert
Vvedensskya pinnatifolia Korov
Xanthoceras sorbifolia Bge
Zanthoxylon bungei Planch

1, also in Turkmenia

Zeravschanica regeliana Korov
Zoegia baldschunanica C Winkl
Zoosteria nottii Hornem

ADDITIONAL SPECIES COMMON TO UZBEKISTAN AND KIRGTZ

Abelia corymbosa Rgl et Schmath
Acinos rotundifolius Pers
Aesculus hippocastatum L
Agrimonia asiatica Juz
Ailanthus altissima Juz
Alliaria petiolata (Bieb) Cevara et Grande
Apluda mutica L
Anthraxon hispidus (Thunb) Makino
Asclepias syrica L
Camphorosma lessingii Litv
Cephalaria syriaca (L) Roem et Schult
Chaenorhinum spicatum Korov
Helerpestis salsiguosa (Pall ex Georgi) Green
Heteropappus canescens (Nees) Novopkr
Incarvillea olgae Rgl
Leptorrhabdos parviflora Bth
Leucanthemum vulgare Lam
Ligustrum vulgare L
Lilium candidum L
Lotus frondosus (Freyn) Kuprian
Maclura pomifera (Raf) CK Schneid
Maloper trifida Cav
Marrubium alternidens Rehb f
Microcephala lamellata (Bge) Pobed
Nicandra physaloides (L) Gaertn
Odontites vulgaris Moench
Oonis antiquorum L
Ostovskia magnifica Rgl
Oxyria digyna (L) Hill
Parapyrum anemonoides (K et K) Ulbo
Petrohagia alpina (Habl) PW Bell & Heywood
Phalaroides arundinacea (L) Rauschert
Phellodendron amurense Rupr
Physocarpus opulifolius (L) max
Picris nuristanica Borum
Pilostemon karategnii (Lipsky) Iljin
Platycladus orientalis (L) Franco
Rhinanthus songaricus (Stern) B Fedts
Schibternanthis longistipitata (Rgl) Nakai
Scirpioides holoschoenus (L) Sojak
Teucrium scordioides Schreb
Trifolium vulgare Nees
Tythostemma alisnoides (Boiss et Bohse) Nevski
Weigela praecox (Lemoine) Bailey

ADDITIONAL SPECIES COMMON TO UZBEKISTAN AND TADJIKSTAN

Albizzia julibrissum Durazz
Ammothamnus lehmanii Bge
Arundo donax L
Astoma galicarpum (Korov) MP et Kljuykov
Bergia ammanoides Roxb
Boehmeria nivea (L) Gaud
Brachiaria eruciformis (Smith) Gris
Brachypodium silvaticum (Huds) PB
Buchingera axillaris Boiss
Buffonia oliveriana Ser
Buglossoides arvensis (L) Johnst
Chalanthus racefolius L
Coleosepala aegacauthoides (Boiss) Boiss
Cressa cretica L
Cutandia memphitica (Spreng) Bth
Cymatocarpus papovii Botsch et Vved
Descurainia sophia (L) Schur
Dipcadi turkestanica Vved
Enneapogon persicis Boiss
Erianthus revennae (L) Beau
Eriochloa succinta (Trin) Kunth
Eriocaulon sieboldianum Sieb et Zucc ex Steud
Eulophia turkestanica (Litv) Schlechter
Exochorda albertii Rgl
Halogeton glomeratus (MB) CAM
Hesperis sibirica L
Laburnum anagyroides Medik
Lithospermum officinale L
Loliolum subulatum (Banks et Soland) Eig
Londesia erianthus F et M

Ludwigia parviflora Roxb
Mertensia dschagastanica Rgl
Neslia apiculata F et M
Paliurus spine-christi Mill
Phacelia tenacetifolia Bth
Phalaris minor Retz
Physocaulis nodosus (L) Koch
Poncirus trifoliata (L) Rafin
Rapistrum rugosum (L) All
Restella alberta (Rgl) Pobed
Salicornia europaea L
Samolus valerandi L
Schtschurowskia meifolia Rgl et Schmath
Spartium junccum L
Spirorrhyncus sabulosus K et K
Tauscheria lasiocarpa F et M
Tetradictylis tenella (Ehr) Litv
Trichodesma incanum (Bge) DC
Tylosperma lignosa (Willd ex Schlecht) Botsch Urtica dioica L
Zeuxine strateumatica (L)Schlecht

Kirgiz

SPECIES NOT COMMON WITH ADJOINING AREAS AND KASHMIR

Achnatherum saposhinikovii (Roshev) Nevski
Amberboa bucharica Iljin
Ammania auriculata Willd
Amelopsis vitifolia (Boiss) Planch
Anchusa italica Retz
Ammopiptanthus nanus (MP) Cheng f
Anthemis tinctoria L
Anthochlamys turcomanica Iljin
Antonia debilis (Bge) Vved
Aphanopleura capillifolia (Rgl et Schmath) Lipsky
Aristida heymannii Rgl
Asterothamnus schischkinii Tamamsch
Betonica foliosa Rupr
Brachanthemum kirghisorum Krasch
Calophaca tianschanica (B Fedt) Boriss
Calystegia sepium (L) R Br
Camelina silvestris Wall
Cardamine impatiens L
Caucalis platycarpos L
Calamintha debilis (Bge) Bth
Crytodiscus didymus (Rgl) Korov
Chamaerhodos sabulosa Bge
Chelidonium majus L
Chrozophora hierosolymitana Spreng
Clausia turkestanica Lipsky
Cleome fimbriata Vicary
Coleoglossum viride (L) Hatm
Comarum salessovianum (Steph) Aschers et Gr
Conioselinum tataricum Hoffm
Crocus alatavicus Rgl et Sem
Cuminum setifolium (Boiss) Pol
Dendrothema morifolium (Ramat) Tzvel
Dictamnus angustifolius Don f ex Sweet
Digitalis purpurea L
Dorema microcarpum Korov
Elscholzia densa Bth
Eminium lehmennii (Bge) OK
Eremosparton songoricum (Litv) Vass
Eriophorum schenchzeri Hoppe
Erophila verna (L) Bess
Exochorda tianschanica Gontsch
Elaeosticta knorringiana (Korov) Korov
Fumaria vaillainti Loissel
Galeopsis bifida Boenn
Gaillonia bruguierei A Rich
Grossularia reclinata (L) Mill
Hierochloa odorata (L) Beau
Hulthemia herberifolia (Pahl) Dum
Hypacanthium echinopifolium (Bornm) Juz
Hypecoum parviflorum K et K
Hyalolaena collina Korov
Hypericum parviflorum K et Vved
Iskandera alaica (Korsch) Botsch et
Korovinia ferganensis Korov Lipsky
Korshinskia olgae (Rgl et Schmath)
Lymeropappus schakaptaricus (N Fedt) Knor et Tammamsch
Lavendula angustifolia Mill
Levisticum officinale Koch Nevski
Littledalea alaica (Korsch) Petr ex
Lobelia erinus L

Lomatogonium carinthiacum (Wulf) Rchb
Lycopus europaeus L
Lychnis chalcedonica L
Lygusticum discolor Ldb
Metastachydium sagittatum (Rgl) Wu et al
Najas madrina L
Neogaillonia brugiveri (A Rich ex DC) Lincz
Nonnea caspica (Willd) G Don
Oenanthe fedtschaukoana K Pol
Omalotheca supina (L) DC
Orthilia secunda (L) House
Padus avium Mill
Paeonia intermedia CAM
Patrinia intermedia (Horn) Roem et Schult
Pennellianthus barbatus Nutt
Petrosimonia sibirica (Pall) Bge
Pilopleura torydyloides (Korov) MP
Platycodon grandiflorus (Jacq) DC
Polemonium caucasicum N Busch
Proterium sanguisorba L
Prangos didyma (Rgl) MP
Pseudoclausia turkestanica (Lipskv) P Vassil
Psederemostachya severtzovii (Herd) MP
Psammanogeton setifolium Boiss
Ramisia secunda (L) Garckle
Reseda luteola L
Rhinopetalum steuanthemum Rgl
Roborowskia mira Betalin
Ruta graveolens L
Salicornia europaea L
Santolina chamaecyparissus L
Schumis arabicus Nees
Schultzia albiflora (K et K) MP
Seminovia dasycarya (Rgl et Schmath) Korov
Sibiraea tianschanica Pojark
Sideritis montana L
Sorbaria olgae Zinseri
Spinacea turkestanica Iljin
Stephanicaryum olgae (B Fedt) MP
Taphrospermum altaicum CAM
Tianschaniella umbellifer B Fedt ex MP
Timouria saposhnikowii Roshev
Trachelanthus korolkovii Lipsky
Tribulus terrestris L
Trichodesma incaum (Bge) DC
Veratum lobelianum Bernh.

ADDITIONAL SPECIES COMMON TO KIRGIZ AND TADJIKSTAN

Anchusa italica Retz
Atragene sibirica L
Chalcanthus renifolius (Boiss) Boiss
Chamaerhodos sabulosa Bge
Cuminum setifolium (Boiss) K-Pol
Grossularia reclinata (L) Mill
Littledalea alaica (Korsch) Petr ex Nevski
Microgynoecium tibeticum HK
Neottia camtkshatea (L) Rchb f
Paenia intermedia CAM
Parligusticum discolor (Ldb) Trichomirov
Parthenicissus quinquefolia (L) Planch
Rorippa palustris (L) Bess
Salicornia europaea L
Schismus arabicus Nees
Securiniga suffruticosa (Pall) Rehd

ADDITIONAL SPECIES COMMON TO KIRGIZ,KHZAKHSTAN AND TADJIKSTAN

Chalcanthus renifolius (Boiss) Boiss
Juncellus serotinus (Rottb) CB Cl
Oxygraphis glacialis (Fisch) Bge
Thylacospermum caespitosum (Camb) Schischk
Thymaelaea passerina (L) Coss et Germ
Trichodesma incanum (Bge) DC

Tadjikstan

SPECIES NOT COMMON WITH ADJACENT REGIONS AND KASHMIR

Albertia paleacea Rgl et Schmath
Alliaria alliaca (Salisb)Brit et Rendle
Anthochlamys tianschanica Iljin ex Aellin
Alposelinium albomarginatum (Schrenk) MP
Apluda inermis Rgl
Arrhenatherum elatius (L) M et K
Arthraxon longsdorfii (Trim) Hochst
Bergenia stracheyi (Hk f et T) Engl

Brothriospermum tenellum (Hornem) F et M
Calymmatium draboides (Korsch) OES
Cannabis ruderalis Janish
Catenularia hedysaroides Botsch
Carya pecan (Marsh) Eng et Gaertn
Celastrus orbiculata Thunb
Cephalopodum badachanicum Korov
Chesniella subulosa Boriss
Citharoloma lehmannii Bge
Cnidiocarpa alaica MP
Clementsia semenovii (Rgl et Herd) Boriss
Comastoma falcatum (Turcz) Toyokuni
Clypeola jonthlaspi L
Cornus mas L
Cryptodiscus dadymus (Rgl) Korov
Cysticorydalis fedtschenkoana (Rgl) Fedde
Danthonia forskohlei R Br
Dendrostellara turkminorum Pobed
Desideria pamirica Suslova
Desmazeria compressa Ovcz et Schisk
Didymopelta turkestanica Rgl et Schmath
Dimorphosciadium gayoides (Rgl et Schmath) MP
Dionysia involucrata Zapr
Dichasianthus subtilissimus (Pop) Ovcz et Junuss
Elatine triandra Schuhr
Evodia hupehensis Doge
Ewersmannia sogdeana Ovcz
Faba vulgaris Moench (cult)
Falcaria polyantha (Korov) MP
Fergania vulgaris Bernh
Fiedleria alpina (Hall) Ovcz
Gladiolus segetum Kar-Gawl
Gomphrena globosa L
Graellsia graellsifolia (Lipsky) Poulter
Gymnocarpium fedtschenkoanum Pojark
Halerpestes sarmentosa (Adams) Kom
Hegemone micrantha
Helosciadium nodiflorum (L) Koch
Heracleum lehmannianum Bge
Heteranthelium piliefrum Hochst
Holoschoenus vulgaris Link
Horanninovia ulicina F et M
Hyalolaena jaxartica Bge
Hyminolyma trichophyllum (Schrenk) Korov
Iskandera hissarica N Busch
Kafirnigania hissarica (Korov) Kamet-Kinz
Keyserbungia mollis (Royle) Boiss
Koeleria gracilis Pers
Ladyginia bucharica Lipsky
Laurocerasus officinalis Roem
Lavatera cashemeriana Cambess
Lawsonia inermis L
Loliolum subulatum (Banks et Sol) Eig
Lomatocarpa steineri (Podlech) MP
Luzula spicata (L) DC
Lychnis coronaria (L) Descr
Maliacurus lanatus (Korsch) Nevski
Mediasia macrophylla (Rgl et Schmath) MP
Meristotropis bucharica (Rgl) Krug
Mogoltavia severzovii (Rgl) Korov
Muehlenbergia huegelii Trin
Muscari bucharicum Rgl
Myrtama elegans (Royle) Ovcz et Kinz.
Narcissus tazella L
Nardurus krausei (Rgl) V Krecz et Bobr
Nasturtium microphyllum Boenn
Nevskiella gracillima (Bge) VKrecz et Vved
Orostachys thyrsiflora Fisch
Orthelia obtusata (Turcz) Jurtz
Padellus mahaleb (L) Vass
Paracynoglossum glochidistum (Wall ex Bth) MP
Parapholis incurva (L) Hubb
Parasitans asiaticus (Korov) MP
Paspalum digitaria
Pelargonium roseum
Platycaria strobilaca Sieb et Zucc
Pleurogynella brachyantha (Cl) IKonn
Prionotrichon pseudoparrya Botsch Vved
Prosopis farcata (Banks et Soland) Macbride
Psathyrostachys kronenburgii (Hack) Nevski
Pseudomelissitus dasycarpus (Ser) Ovcz Rassul et Kuiz
Pseudomertensia rosulata (Ovcz Czak) Ovcz et Czak
Psilurus aristatus (L) Dnv-Jouve
Pueraria hirsuta (Thunb) Schneid
Pyrola rotundifolia L

Sageretia latevirens (Kom) Gontsch
Sanguisorba alpina Bge
Schinus dependens Ortega
Schulzia crinita (Pall) Spreng
Siella erecta (Huds) MP
Sium sisaroideum DC
Smelowskia calycina (Steph) CAM
Sophora japonica L
Sphaerophysa salsula (Pall) DC
Tetrataenium olgae (Rgl est Schmath) Manden
Thlaspidium thlaspi (Lipskv) Rassul
Tillaea aquatica L
Trachylanthus hissaricus Lipsky
Trichophorum pumillum (Vahl) Schinz et Thell
Trollius dschungaricus Rgl
Verbena officinalis L
Vicatia conifolia DC
Vincetoxicum darvasicum B Fedt
Vitex agnus-castus L
Zanichellia palustris L
Zosima korovinii MP

SPECIES COMMON TO TADJIKSTAN, UZBEKISTAN AND KIRGIZ

Apera interrupta (L) PB
Apium graveolens L
Berteroa incana (L) DC
Biebersteinia multifida DC
Boissiera squarrosa (Soland) Nevski
Callianthemum altavicum Freyn
Camphorosma lessingii Litv
Didymophysa fedtschenkoana Rgl
Glyceria plicata Fries
Helianthemum songoricum Schrenk
Imperata cylindrica (L) PB
Microsisymbrium minutiflorum (Hk f &T)OES
Milium vernale MB
Nanophyton erianceum (Pall) Bge
Panicum miliaceum L
Pastinaca sativa L
Phellodendron amurense Rupr
Phragnites communis Trin
Polygala hybrida DC
Ptelea trifoliata L
Sagina saginioides (L) Dall-Torre
Sagittaria trifoliata L
Schoenus nigricans L

SPECIES COMMON TO TADJIKSTAN, UZBEKISTAN, KIRGIZ AND KAZAKHSTAN

Agrostema grifithii L
Amorpha furticosa L
Arceuthobium oxycedri (DC)MB
Armoraci rusticana Gaertn, Mey et Schreb
Arum borolovii Rgl
Asperugo procumbens L
Cladium mariscus (L)Pohl
Cotinus coggygria Scop
Cynomorum songoricum Rupr
Diptychocarpus strictus (Fisch)Trautv
Eremodaucus lehmannii Bge
Glaux maritima L
Halimodendron holodendron (Pall)Vass
Hippophae rhamnoides L
Ilingia regelii (Bge)Korov
Kirilovia eriantha Bge
Lachnoloma lehmannii Bge
Leptaleum filifolium (Willd)DC
Litwinowia tenuissima (Pall)Woron
Lycopsis orientalis L
Meniocus linifolius (Steph)DC
Nanophyton erinaceum (Pall)Bge
Petroselinium crispum (Mill)Hill
Psoralia drupacea Bge
Saccharum spontaneum L
Salicornia europaea L
Sphaerophysa salsula (Pall)DC
Trichodesma incaum (Bge)DC
Valisneria spiralis L
Velezia rigida L
Zizyphus jujuba Mill

Turkmenia

GENERA WITH SINGLE SPECIES

AMARANTHACEAE

Gomphrena globosa L.

NYCTAGINACEAE

Mirabilis jalapa L.

AIZOACEAE

Aizoon hispanicum L.

POTULACACEAE

Potulaca oleracea L.

CARYOPHYLLACEAE

Melachium aquaticum Fries
Queria hispanica L.
Telephium orientale Boiss
Agrostemma githago L.
Vaccaria segetalis (Neck)Garcke
Velezia rigida L.
Velezia rigida L.

RANUNCULACEAE

Eranthus longistipitata Rgl
Paraquilegia karelinia (Drumm)Nevski
Aquilegia vicaria Nvski
Myosurus minimus L.
Ficaria fascicularis C Koch

BERBERIDACEAE

Leontice ewersmannii Bge
Bongardia chrysogonum (L)Boiss
Berberis turcomanica Karel

CRUCIFERAE

Alliaria alliacea (Salisb)Rendlex Britt
Microsisymbrium minutiflorum (HK et Thom)OES
Alyssopsis kotschyi Boiss
Thellungiella salsuginea (Pall)OES
Drabopsis verna C Koch
Descurainia sophia (L) Webb
Syrenia siliculosa (MB)Andrz
Roripa palustris (Leyss)Bess
Nasturtium officinale (L)R Br
Cardamine hirsuta L.
Turritis glabra L.
Graellsia saxifragifolia (DC) Boiss
Chartoloma platycarpum Bge
Spirorrhynchus sabulosus K et K
Clausia turkestanica Lipsky
Cryptospora falcata K et K
Tetracmidion bucharicum Korsh
Leptaleum filifolium (Willd)DC
Streptoloma desertorum Bge
Diptychocarpus strictus (Fisch)Trautv
Chorispora tenella (Pall)DC
Strerigmostemum tomentosum MB
Anchonium ramosissimum OES
Trichochiton inconspicuum Kom
Euclidium syriacum R Br
Litnovia tenuissima (Pall)N Busch
Octoceros lehmannianum Bge
Lachnoloma lehmanni Bge
Fibigia suffruticosa Sweet
Meniocus linifolius (Steph)DC
Buchingera axillaris Boiss
Erophila verna (L)Bess
Sinapsis arvensis L
Eruca sativa Lam
Pistrum rugosum (L) All
Calepina irregularis (Azso) Thell
Spryginia winkleri (Rgl) M Pop
Chalcanthus tuberosus Kom
Cardaria draba (L) Desv.
Hymenolobus procumbens (L) Nutt.
Carpoceras stenocarpum Boiss
Peltaria turkmena Lipsky
Nesslia apiculata Fisch et Mey
Capsella bursa-pastoris (L) Medd.

PRIMULACEAE

Primula fedtschenkoi Rgl
Androsace turczaninovii Freyn
Samolus valerandi L
Glaux maritima L
Anagalis arvensis L

PLUMBAGINACEAE

Plumbago europaea L
Chaetolimon setiferum (Bge)Lincz
Cephalorrizum turcomani M Pop

OLEACEAE

Olea europaea L
Ligustrum vulgare L

Jasminum fruticans L
GENTIANACEAE
Gentiana olivieri Griseb
APOCYANACEAE
Vinca major L
Trachomitum scabrum (Russau) Pobed
Nerium olenderi L
ASCLEPIADACEAE
Periploca gracea L
Antitoxicum pumilum (Dcne) Pobed (=*Vincetoxicum* Moench)
Cynanchum sibiricum Willd.
CONVOLVULCEAE
Cressa cretica L
Ipoemia purpurea (L) Roth
Calystegia sepium (L) R Br
BORAGINACEAE
Suchtelenia acanthocarpa Kar
Trichodesma incanum (Bge)DC
Coccinia crassifolia (Vent)C Koch
Cynoglossum pictum Ait
Lindelofia macrostyla (Bge)M Pop
Bilegnum bungei (Boiss)Brand
Asperugo procumbens L
Gastrocotyl hispida (Forsk)Bge
Borago officinale L
Anchusa italica Retz
Lycopsis orientalis L.
VERBINACEAE
Lippia nodiflora (L) Rich
LABIATAE
Ajuga pseudochia Schost
Sideritis montana L.
Lophanthus lipskyanus IK-Gal et Nevski
Kudrjaschevia jacubi (Lipsky) Pojark
Dracocephalum subcapitatum (Ktza)Lipsky
Hypogomphia turkestana Bge
Prunella vulgaris L
Leonurus turkestanicus Krecz et Kupr
Molucella laevis L
Thuspeinantha persica (Boiss) Briq
Chamaesphacos ilicifolius Schernk
Melissa officinalis L
Clinopodium intergerrimum Boiss
Acinos graveolens (MB) Link
Origanum kopetdaghensis Boriss
Majorana hortensis (L) Moench
Thymus transcaspicis K et K
Ocimum basilicum L
SOLANACEAE
Nicandra physalodies (L) Geartn
Atropa komarovii Blinovsky et shalyt
Mandragora turcomanica O.Mizgir
Datura stramonium L.
SCROPHULARIACEAE
Celsia heterophylla Desf
Kicksia elatina (L) Dum
Chaenorrhinum spicatum Kor
Paulownia tomentosa Stend
Dodartia orientalis L
Triaenophora bucharica B Fedt
Leptorhabdos parviflora Bth
Euphrasia tatarica Fisch
Parentucellia flaviflora (Boiss) Nevski
Ortantha aucheri (Boiss) Wettst
BIGNONIACEAE
Campsis radicans (L) Seem
Chilopsis linearis (Cav) DC
LENTIBULARIACEAAE
Utricularia vulgaris L.
RUBIACEAE
Sherardia arvensis L.
Callipeltis cucullaria (L)Stev
CAPRIFOLIACEAE
Sambucus ebulis L.
CAMPANULACEAE
Campanula incanescens Boiss
Asyneuma regelii (Trautv)Bornm

AFGHANISTAN

Linchevsky and Prozorovsky (1949) recognised four geobotanical regions in Afghanistan as follows (Fig 16).

I.South Turkestan Ephemeral Province: 30% of total area of Afghanistan. Geobotanical regions: 1) Amu-darya lowland, 2) Cisparapamiz low mountain, 6) West Badakhasan high mountain, 7)East Badakhasan high mountain.

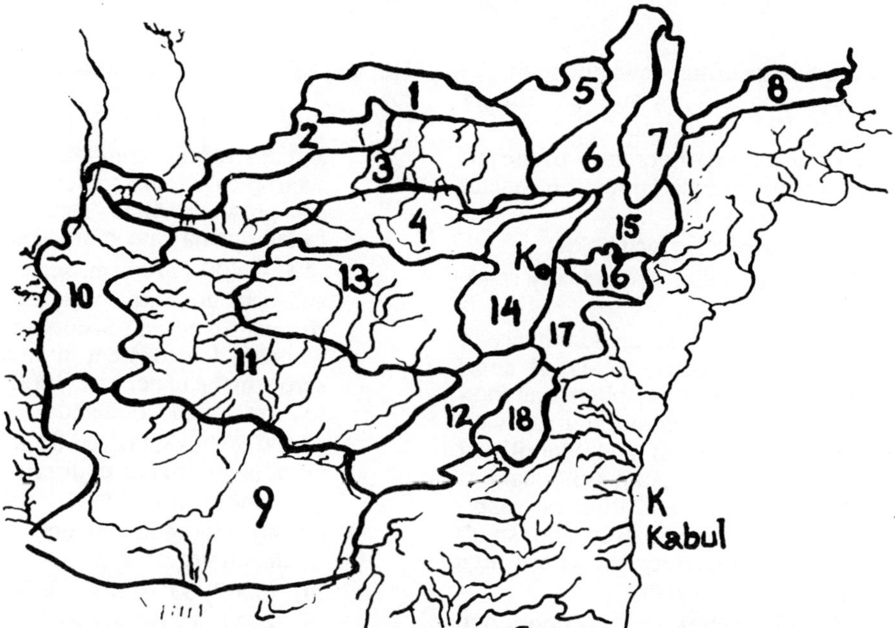

Fig. 16 : The geobotanical regions of Afghanistan (for explanation see the text, cf Linchevsky & Prozorovsky,1949)

Characteristics: meadow like ephemeroid and ephemeral vegetation;in intermediate mountain zones trees and shrubs, steppe at higher altitudes.

The altitudinal profile is given below.

1. up to 900 (1100)m. Low herbage ephemeroid zone:
   (a) up to 900-600m, ephemeral-sedge-medow-grass vegetation.
   (b) up to 900-1100m, coarse herb-sedge-meadow-grass vegetation and Pistachio communities.
2. 900 (1100)-1800 (2000)m. Medium herb ephemeroid vegetation and Pistachio communities and scrub.
3. 1800 (2000)-2500 m (in the west)- 3000m (in the east) tree and shrub vegetation,sparse juniper communites and scrub in the east (Western Badkhashan) + maple and walnut woods.
4. 2500 (3000) - 4000 m. High mountain vegetation:
   (a) 2500 (3000)-3500 m, subalpine zone, preponderance of drought resisting vegetation on sandy soils (steppes, tragacanth) Cousinia and Ferula communities
   (b) 3500-4000 m, alpine zone, preponderance of drought resisting vegetation (steppes tragacanth and Cousinia communities) and low herbage meadows.
5. Above 4000 m, no vegetation.

II. Central Asiatic Desert Province: extreme northeastern Afghanistan, 3% of the territory. Geobotanical region, 8) Wakhan high mountain.

Characteristics: desert semifruticose and steppe vegetation in alpine zone,low herbage Cobresia meadows.

The altitudinal profile is as follows:

1. 2000-3700 m, semifruticose vegetation, preponderance of worm wood and *teresken* communities.
2. 3700-5000 m, high mountain vegetation:
   a) 3700-4000 m, subalpine zone, high mountain steppe.
   b) above 4000 m, alpine zone, low herbage, chiefly Cobresia and Atropis meadows.

III. Afghano-Iranian desert province, southern part of Afghnistan,60% of whole country. Geobotanical regions:9) Seistoan-Registan lowland. 10) west Afghanistan mountain and lowland, 11) south Afghanistan low mountain. 12) Ab-i. Istadah mountain and lowland, 13) Hazara high mountain, 14) Kabul mountain.

Characteristics: preponderance of desert semifruticulose and tragacanth-fruticulose vegetation.

The altitudinal profile is given below.

1. up to 3200 m, desert (principally semi-fruticulose):
   (a) up to 1000-1200 m, predominance of earthy and rubbly soils of semi fruticulose vegetation, plains and foothills,(pure worm-wood and salt-wort communities),plains and foothills of southern Afghanistan with desert-semifruticulose vegetation on sands, with meadows and tree and shrub vegetation in the river valleys with high level of subsoil water:
   (b) 1000(1200)-2400 m, preponderant ephemeral-worm wood and feather grass- worm wood desert communities (on rubbly slopes of medium-and-low altitude mountains),also sparse Pistachio communities here and there; tragacanth shrublets (chiefly admixture in worm- wood communities) + ? thicklets of drought resistant shrubs:
   (c) 2400-3200 m, preponderance of tragacanth-worm wood and feather-grass wood desert communities (lower part of the high mountain tract).
2. 3200-4400 m, high mountain (principally herbaceous) vegetation:
   (a) 3200-3800 m, subalpine preponderant fescue steppes, + considerable scattering of Cousinia and tragacanth communities (on stony slopes):
   (b) 3800 - 4400 m, alpine zone, preponderance of low mixed herbage and grass meadows. Cousinia communities and fescue steppes.

IV. Indo-Himalayan Forest province: southeastern Afghanistan,10% of the total area. Geobotanical regions: 15) Nuristan high mountain, 16) Dzhelalabad, 17) Safid Kuh mountain, 18) Upper Gumal mountain plain.

Characteristics: Preponderant natural plant covering of trees and shrub vegetation of Indo-Himalayan type. The altitudinal profile is as follows:

1. (750)-1200 (1500) m, subtropical steppes and desert grass communities + thickets of subtropical shrubs, sometimes considerable low growing palms (nannorhops) may be called savana type in broad sense.
2. 1200 - 2400 (2700) m, scrub communities and deciduous (in the upper part of coniferous-decidous) woodland; basic oak woods,+ development of undergrowth and scrub; in the upper part (1700 m). + conifers to form mixed woodland.
3. (2400)2700-3000(3350) m, coniferous woodlands, basic coniferous woodlands with small proportion of deciduous species.
4. (3000) 3350-4000 m and upwards, high mountain vegetation:
   (a) (3000) 3350-3700 (4000)m, subalpine zone, basic scrub communities of procumbent junipers, willows and rhododendrons of xerophil-mixed-herbage (Cosinia) communities.and subalpine grass meadows;
   (b) (3700) 4000 m and upwards, alpine, basic alpine Allium grass-Cobresia meadows, xerophil-mixed herbage communities and sparse vegetation of rocks and stony screes and wastes.

Natural vegetation

Freitag (1971) summarized the role of precipitation on the vegetation of Afghanistan as follows:

1) The rainy season lengthens from south to north, stopping in south in early April and in north after May.
2) The south is more prone to run-off to heavy rains, the north is interesting in that the places with similar mean precipitation have more rainy and cloudy days.
3) The monsoon rarely extends to near about Faizabad and Kandahar. Few places do receive regular summer rain.

He concluded that such conditions favour luxuriant vegetation in the north. Consequent upon this the ecological situation depicts that in areas with remarkably high precipitation the woodlands flourish and where climate is alike throughout the year, Cedrus forests show great development but in areas without summer rains "but even higher precipitation" only Amygdalus woodands grow.

The main vegetational types distinguished by Freitag are the following.

I. Semi-desert vegetation (prec 250-300 mm,trees absent).

a) Calligonum-Aristida communities :

i) drier parts : *Haloxylon persicum, Calligonum molle, C. turkestanicum, C. camosum, C.intertextum, C.amoenum.*

ii) shrubs (and small trees) : *Aristida pennuata, A.kerelinii Cyperus conglomeratus* (perennial); *Agriophyllum minus, A.latifolium, Euphorbia chaeirolepis, Cutandia memphitica, Girgensohnia minima*(annuals).

b) Marginal areas (prec higher) : *Carexhysodes, Sophora pachyclada, Ammothamnus lehmannii* (mobile sand fixers), *Stellaria lessertii, Convolvulus hemadae, Artemisia scoparia, Eremostachya regeliana, Eminium alberti* (desert species).

II. Chenopodiaceous communities of extreme semi-desert (prec less than 150 mm).

a) Haloxylon salicornicum community (southwest Afghanistan), *Haloxylon salicornicum, Salsola arbuscula, Seidlitzia rosmarinus, Anabasis setifera, Ephedra scoparia* (dominants)

b) Widespread types

i) Salsola-Salvia trichocalycina community: *Salsola subaphylla, Acatholimon longiflorum, Salsola canascens* (dwarf shrubs. numerous).

ii) Zygophyllum tetrapterum- Cousinia arida community (west of Afghnistan);

iii) Reaumuria stocksii community (rocky soils, southwest Afghanistan).

iv) Artemisia-Ferula community (west of Af-ghanistan): *Ferula assafoetida, F.badrakema, Dorema aitchisonii.*

v) Seidlitzia rosmarianus-Salsola gemmascens community (narrow belt near Amu-Darya) *Chenopoda* (dominant).

III. Amygdalus shrubland communities (transition between extreme semi-desert and open woodland, 150-250 mm). Scattered thorny bushes: dwarf shrubs dominant, geophytes and annuals in interspaces: Artemisia cf eburnea community of west and southwest Afghanistan replaced by *A. erioclada* in Hilmand region and by A. brahuica communities in isolated regions of west and east Afghanistan.

IV. Ephemeral semi-desert communities of loess-belt (150-300 mm) in north Afghanistan .

a) dominant hemicryptophytes: *Poa bulbosa, Cousinia alata.*

b) geophytes: *Carex stenophylla, Marendera robusta, Gagea* several (sev sps), *Vulpia* luxriant annuals, *Bromus* sev sps *Vulpia myurus, Aegilops tauschii, A. triuncialis, Eremopyron distans, Diarthron vesiculosus, Trigonella grandiflora, Aphanopleura leptoclada,* etc.

c) dwarf shrubs and tree shrubs: rare to absent.

V. subtropically influenced shrub-and grassland communities:(150-300 mm, hot basin of Jalalabad and surroundings):

a) Zizyphus nummularia community (up to 750 mm): *Zizyphus nummularia, Rhazya stricta, Acacia modesta, Salvia aegyptiaca, Eleusine flagellaris,* etc.

b) Salvia cabulica-Pistacia khinjuk community: *Periploca aphylla, Pistacia khinjuk, Amygdalus kuramica* ssp *aitchisonii, Ephedra intermedia, E. ciliata, Salvia cabulica.*

VI. Open deciduous woodlands (250-300 mm) semidesertic: Amygdalus shrublands and ephemeral communities are replaced by extreme xeromorphic woodlands, Chenopodiaceae and Zygophyllaceae are absent.

a) Pistacia atlantica community (1150- 1800 m; 250-400 mm):

i) tree layer, *Pistacia atlantica* and *P.khi juk* (at lower altitudes).

ii) Shrub layer: *Amygdalus* sev sps, *Convolvulus leiocalycinus, Astragalus mokurense, Ephedra* sps, etc.

iii) dwarf shrubs: *Cousinia stocksii-Astragalus* sev sps , *Gaillonia* sps *Stachys trinervis,* etc; herbaceous perennials, annuals and geophytes.

b) Pistacia vera community (600-1500 m; 300-1500 mm), northern Afghanistan.

i) trees (2.5 m tall): *Pistacia vera, Amygdalus bucharica*

ii) shrubs : *Amygdalus spinossima, Cerasus bifrons.*

iii) sub-shrubs : *Artemisia tenuisecta, Acatholimon cabulicum,* etc;

iv) hemicryptophytes: *Cousinia reddena. C.alata, C.umbrosa, Phlomis bucharica, Salvia pterocalyx.*

v) geophytes : *Carex stenophylla, Eremurus bucharicus, E. olgae, E. regellii, Bongardia chrysogonum, Eranthis longistipitata,* etc.

vi) annuals : *Aegilops cassa, Bromus* sev sps. *Taeniatherum Crinitum, Hypoghomphia turkestanica,* etc.

c) Amygdalus communities: between Pistacia atlantica belt and subalpine region 2000-2800 m, more than 400 mm).

i) tree layer: Amygdalus cf communities: *A karamica, Fraxinus zanthoxyloides:*

ii) shrubs : many species of *Cotoneaster* and *Rosa, Cerasus bifrons, C.verucosus, Colutea persica, Ephedra intermidia,* etc.

iii) dwarf shrubs: many species of *Artemisia, Astragalus, Acantholimon, Perovskia atriplicifolia,* etc.

iv) hemicryoptophytes: *Cousinia racemosa, C. ghazianiansis,* and other sps. *Salvia bucharica, Leucopoa karatavica,* etc.

v) geophytes : various species of *Eremurus, Allium, Gagea, Tulipa, Corydalis,* etc (rich in endemics).

VII. Evergreen sclerophyllous forests and woodlands:

a) Reptonia buxifolia community at lower altitudes adjoining semidesertic Zizyphus nummularia community; at higher altitudes Quercus baloot forests; (800-1300 m).

i) small trees: *Reptonia buxifolia, Olea ferruginea, Acacia modesta, Pistacia khinjuk.*

ii) shrubs: *Sageretia brandrethiana, Gymnosporia royleana, Dodonaea viscosa, Nannorrhops ritchieana,* etc.

iii) Perennial grasses: *Cymbopogon parkeri, Herteropogon contortus, Apluda mutica, Hyarrhaenia hirta.*

b) Quercus baloot community (1300-2000 m; 350-400 mm).

i) typical climbers : *Lonicera griffihtii, Rosa brunonii.*

ii) shrubs : *Daphnaea angustifolia, Sophora griffithii, Plectranthus rugosus, Spiraea brahuica,* etc.

iii) herbaceous ground: *Lepechiniella microcarpa, Oryzopsis* sps.

iv) clearings: *Nepeta pinctorium,* etc.

c) Quercus dilatata and Q. semicarpifolia community (summer rains heavy):

i) *Q. dilatata* replaces *Q.baloot* at 1900-2000 m, associates: trees and shrubs: *Juglans regia, Acer turkestanicus, Celtis caucasica, Pyrus pashia, Parratiopsis jacquemontiana, Indigofera gerardiana, Viburnum cotinifolium, Cotoneaster rosea.*

herbs: *Teucrium royleana, Dioscorea del-toides, Brachypodium silvaticum, Festuca gigantea, Rumex nepalensis, Artemisia roxburghiana, Dryopteris spinulosa, Aquilegia afghanica,* etc

ii) Q. semicarpifolia community, trees and shrubs: *Taxus wallichiana, Viburnum cotinifolium, Syringia emodi, Euonymus fimbriatus, Lonicera quinquelocularis* etc.

GROUND VEGETATION : *Nepeta erecta, Salvia rhytidea, Calamintha umbrosa, Podophyllum peltatum, Polygonum amplexifolius, Carex cardiolepis,* etc.

VIII. Evergreen coniferous forests and woodlands (mainly summer rain areas in the east).

a) Pinus gerardiana community (2100-2500 m, between Q. baloot forest and Cedrus community).

i) dwarf shrubs: species of *Artemisia, Astragalus, Acantholimon,* etc.

ii) Cedrus deodara community, 2500-3100 m; 450-650 mm. *Cedrus deodara, Juniperus seravschanica* in areas with lesser rainfall.

a) better water supplied areas: *Pinus wallichiana.*

b) shrubs poorly represented.

c) open areas: *Salvia nubicola, Saussurea, amplifolia, Phlomis cashmeriana, Nepeta pinetorum,* etc.

iii) Picea smithiana- Abies wallichiana com- munity (more than 800 mm, 2900-3300 m).

a) occasional: *Pinus wallichiana, Quercus semicarpifolia.*

b) characteristic in shrub layer: *Patrya aitchisonii, Rubus irritans, Androsace himalaica.*

c) hemicryptophytes (dominant), *Polygonum amplexicaule, Rumex nepalensis, Fragaria nubicola, Pseudomertensia edelbergii, Cicerbita* sps.

d) geophytes: *Polygonum geminiflorum, Lilium polyphyllum, Hebenaria aitchisonii.*

iv) Juniperus community

a) *Juniperus excelsa,* northwest from Paropamirus to Band-e-Turkestan.

b) *J. semiglobosa,* northeast higher altitudes.

c) Juniperus woodlands, east Afghanistan, *J.semiglobosa* 3000-3500 m, *J. seravschanica* lower altitudes. *Cercis griffithiana, Acer semenovii, Crataegus songarica, C. tukestanica, Pyrus karschinskyi, Malus turkmenorum, Ephedra equisetina, Lonicera nmmularifolia,* species of *Cotoneaster, Rosa, Cerasus* (in the north), species of *Cousinia, Artemisia, Astragalus, Acantholimon* associated with those of *Eremurus, Allium, Tulipa, Gagea, Corydalis, Scaliger,* etc.

IX. Subalpine thickets and cushion shrublands (high summer rainfall, east Afghanistan, 3300-4000 m).

i) Juniperus squamata community: *Lonicera obovata, L.webbiana,* etc, herbs as in VIII(C). Picea-Abies forests.

ii) associated *Ribes alpestre, R.villosum.*

iii) Juniperus nana community:

a) dwarf cushion shrubs: *Onobrychis spinosissimum, Astragalus strobiliferus, A. ajfredi,* numerous hemicryptophytes and geophytes.

b) cushion shrubland community: species of *Onobrychis, Astragalus, Acantholimon, Cousinia, Artemisia, Ephedra,* many endemics, *Astragalus cuneifolius,* few species of *Rosa, Rhamnus*(shrubs) species of *Cousinia, Astragalus, Trigonella* etc. (herbs), *Gagea* cf *hissarica, G. setifolia, Eremurus kaulfmannii, Iris porphyrochrysa* (geophytes).

X. Alpine Vegetation: Himalayan type and Pamir Alai types, snow line north facing 4000 m, south facing 5300 - 5400 m.

Elements in the Afghan flora (Kitamura, 1960).

1. Cosmopolitan : most of these have got naturalised through human agency from ancient times: *Poa annua, Taraxacum officinale, Sonchus oleraceous.*
2. NORTHERN HEMISPHERE : plants including those which are distributed throughout the northern hemisphere: *Rorippa islandica, Caltha palustris, Polygonum viviparum.*
3. PANTROPICAL : including mostly such plants as have been naturalised by the equitorial migration since 16th century.
4. EURASIAN : including a wide array of plants which are widely distributed from western Europe to eastern Asia, being usually either monotropical or often represented by several vicarious subspecies: *Picris hieracioides, Solidago virgaurea, Prunella vulgaris, Thymus serpyllum, Lotus corniculatus.*
5. WESTERN ELEMENTS : many of the western Eurasian plant species are found in Afghanistan, some extend even to Africa.
6. CENTRAL ASIAN ELEMENTS : a large number of species are met with ubiquitously in Central Asia (including Afghanistan, Iran and the Russian Turkestan) extending to Mongolia. Remarkable instances are *Lactuca tatarica* and *Centaurea ripens.* Since these show no signs of variation they may have been distributed by the nomads or caravans from Central Asia to Mongolia. This group includes mostly the alpine plants having wide distribution along the moun-

tain chains over the Altai. Tien Shan, Pamir, Hindukush, the Himalaya, Tibet and southwest China; some species even extend to Taiwan.

7. EASTERN ELEMENTS : many of the cold and temperate eastern Himalayan species (Sino-Japanese) occur in east Afghanistan and seem to have migrated to the west along the Himalaya and Hindukush mts which have probably provided corridors for this migration.

THE KASHMIR-HIMALAYA REGION

"The Himalaya arose between the Indian peninsula and Central Asia as a young system of high mountains from west to east" (Meusel, 1971). They connect the main mts of the Near East and Central Asia with those of east Asia. From a floristic point of view some consider Himalaya to be a bribge between the holarctic and extra tropical Eurasian flora. Some consider the range as the eastern extension of the Sino-Japanese region. At the western-end Himalaya touch the Irano-Turanian region, a part some consider in a widest sense of the Mediterranean territory. With regards to their flora the Himalayan ranges are often regarded as a bridge between these two important centres of origin of the holarctic and extra tropical Eurasian floras (Meusel, 1971).

The geological finds from Kashmir and Hazara (Pakistan) have shown that land connections prevailed in this area during the Silurian/Devonian period when Psilophytales flourished in many parts of the world. This ancient flora may occur in many brackish or freshwater deposits of this age in this region since this group has been discovered in the Silurian rocks of Australia as well as NE Europe and the Devonian of China. In fact the Ordovian and Silurian shales of Spiti have revealed traces of plants several of which exhibit forked habit of the Psilophytales but lack vascular tissue, or of the large terminal sporangia with cutinized spores characteristic of Psilophytales (Sahni,1938).

The geological data available on Kashmir, the Salt Range and Central India in consistence with palaeobotanical facts reveal that India was under ice during the Upper Carboniferous and in any case not later than the Lower Permian. However as regards the relations of the flora to the climate Sahni, (1938) held that in India and South Africa the "Glossopteris flora had already come in before the land was free of ice and that it even included some hardy remnants of the pre-glacial vegetation". One can not say in what manner such a wide spread glaciation may have affected the pre-existing flora though it could not wholly exterminate whole or all traces of life even over the whole Gondwanaland at the same time": the nitches with micro-climate of their own could have assisted survival of some species through a glacial period and it is possible that some taxa of the Gondwana flora could have been lineal descendants of the pre-glacial flora (Sahni,1937). Thus early phases of the Glossopteris flora overland lapped with the later end of the glaciation. Consider the vegetation flourishing around and in vicinity of glaciers and ice shales even today as in Greenland, Norway, New Zealand etc and even in Himalaya (cf Sahni 1937). However , the Gondwana flora was not a "glacial flora like that of modern flora of the arctic region " (Sahni. 1937).

Following the Great Ice Age, the characteristic flora was distinct from the contemporary flora of the northern hemisphere and also from the pre-glacial vegetation in the south . Regarding the new form of life (animals and plants) that appeared simultaneously over the whole of Gondwanaland one can only guess. These could have come from remote part of the Antarctic Continent itself orwere essentially an indigenous element of southern origin. In fact Sahni has suggested that the Glossopteris flora must have its roots somewhere in the past into the preglacial flora.

In connection with the climatic relationships of the Glossopteris flora a discovery of unusual interest has been in the Salt Range at Lathwari at a level only 6-8 m above the glacial bed where Gee (in Sahni, 1938) had collected Glossopteris, Gangamopteris and other lower Gondwana plants. However till recently the Gondwana rocks reported from the Himalaya lay to the south of the Indus suture zone (i.e. supposed line of collision between the Indian and Eurasian plates (Gansser, 1966). while Sharma et al (1980) extended the range to north of the Indus Suture Zone. Considering the observations of Stoklin (1968), Klootwizk (1979) and Andriexu and Brunel (1977) they suggested that "three geotectonic regions can be demarcated: the northern Eurasian block, the southern Indian subcontinental block and in-between a mixed zone of Iran-Afghanistan-Hindukush- Karakoram-NE

Ladakh-W Tibet which has blocks of both Gondwanic and Siberian affinities".

From the study of Palaeozoic flora of Siberia and suspected existance of Gondwanic elements some degree of affinity must have existed, notwithstanding the barrier of the Tethys Sea which separated the two botanical provinces. Possibly some means of communication for land plants did exist across the Tethys. Zalessky believed on Palaeo botanical grounds that either the marine barrier was bridged over by an isthmus or a dense archipelago existed and served as a stepping stone for migration of Gongwana plants to the north. Graban stated that it was possible that "this northward migration of the Gondwana flora occurred during Carboniferous (Pre-Permian) time, though so far as we know there were no direct land connections". Wadia in Kashmir and Mushketov in north of the Pamir plateau have proved that during Late Palaeozoic times there was actually a series of land patches dotted over the Tethyls in this region; and these islands may well have served as a means of intermigration between the two floras (Sahni, 1938). In fact it is now agreed that the edge of the Gondwanaland arc is roughly parallel to the main Himalayan arc as a result of the Late Tertiary building movements".

The period of maximum movement in the Tertiary era too appears to have been in Early Meiocene times and regression of the bordering seas and uplift of the border ranges continued into the Pleistocene epoch. The climatc of the Indian region was much cooler then and the Himalayan glaciers had attained their greatest spread then, although volcanic eruptions broke out locally in upper Burma.

There appears to have been a large southern continent Gondwanaland in Palaeozoic times which broke up and drifted northward as fragments Indian and Africa towards and over equator. There also appears to have been a northward continent Angaraland which drifted southwestward as Europe-Asia against the northern end of India and Africa .. The Indian peninsula has been severely fractured in various directions by the thrust from the Asiatic mass, and local irregularities such as the rise of the Assam range and the mountains of Burma and the northwestern region of India have appeared".

The present flora of Kashmir valley is relatively poor due to the intense changes in climate and vegetation consequent upon the rapid uplifting in geologically recent times of the outer Himalayan chain. Many of the pleurizonal species of Kashmir scrub eg *Verbascum thapsus, Silene inflata, Achillea millefolium* are synanthropical immigrants; and some of the characteristic shrubs like *Indigofera, Plectranthus* are outposts of the subtropical semiarid vegetation; shrubs like *Rosa webbiana, Cotoneaster* sps, *Berberis* sps represent contact with the extra-tropical semiarid flora of Central Asia. In fact Kashmir scrub is very closely related to the semiarid and arid vegetation of Central Asia as well as of such perennial herbs as *Nepeta, Astragalus, Trigonella, Salvia* etc (Meusel, 1971).

Present flora of Kashmir

Floristically the state of Jammu and Kashmir is divided into three floristic regions:

(a) alpine desert of Ladakh,

(b) predominantly dry temperate Kashmir and wet temperate Chenab valley, and

(c) subtropical Jammu.

In Kashmir the synopsis of the flora is as follows:

	Genera	Species
Dicotyledons	799 genera	2393 species
Monocotyledons ..	232	706
Gymnosperms ...	10	31

Of the total of about 3130 species collected so far, 44 families are represented by a single genus in Kashmir and their distribution is interesting. Some of them include Trapaceae (*Trapa*), Datiscaceae (*Datisca*), Linaceae (*Linum*), Pyrolaceae (*Pyrola*).Punicaceae (*Punica*), Budlejaceae (*Budleja*), Parnassiaceae (*Parnassia*), Adoxaceae (*Adoxa*), etc. The families with largest number of taxa are the following.

Family	Genera	Species
Compositae	94	395
Gramineae	100	250
Papilionaceae	42	163
Cyperaceae	9	148
Cruciferae	52	141
Rosaccae	29	136
Labiatae	33	127
Scrophulariaceae	26	95
Umbelliferae	44	87
Caryophyllaceae	10	83

The genera with largest number of species in Kashmir include: *Carex 74, (Taraxacum 70),*

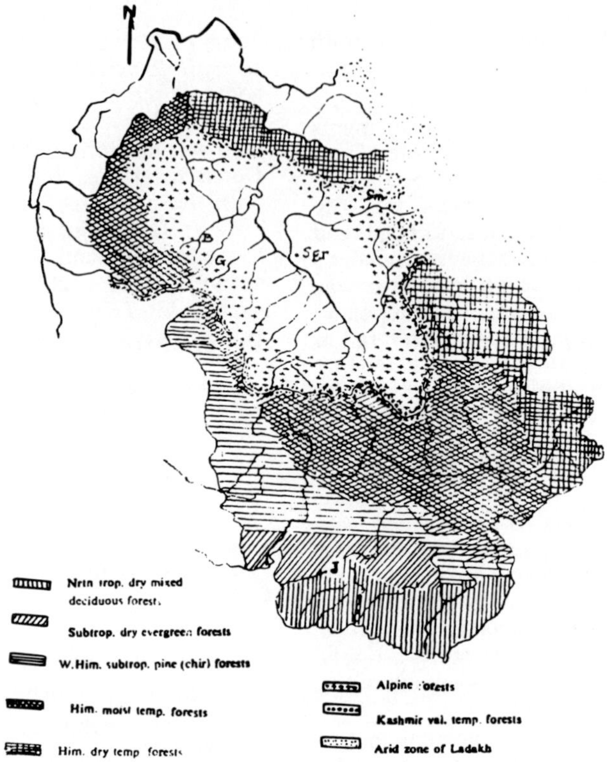

Fig 17. Forest types of Jammu and Kashmir State (cf. Vishnu-Mittre, 1967). Sgr, Srinagar B. Baramulla G.Gulmarg J. Jammu.

Astragalus 55, *Potentilla* 45, *Polygonum* 41, *Nepeta* 37, *Gentiana* 37, *Saussurea* 35, *Artemisia* 32, *Ranunculus* 28, *Cyperus* 28, *Pedicularis* 27, *Veronica* 26, *Corydalis* 25, *Primula* 21, *Draba* 20, *Gerasium* 19.

The forests of Jammu and Kashmir (Fig. 17) fall under the following five main types:

1. Subtropical. 1200-1400 m
   - (a) Subtropical chir pine: *P. roxburghii* scattered + scrub.
   - (b) Subtropical Olea: *O. cuspidata.*
2. Temperate (dry) 1600-2100 m
   - (a) *P. griffithii*, clayey alluvium, drier soils, N/NW
   - (b) *Cedrus deodara*, loam, N.
3. Montane 2100-2300 m
   - (a) mixed coniferous: *Cedrus deodara, P. griffithii, Abies pindrow, Picea smithiana.*
     - (i) in subtypes *Cedurs, Pinus*, or *Abies* as dominant.
     - (ii) *Picea* absent under dominance of *Abies.*
   - (b) 1800-2500 m: deciduous : *Juglans regia, Aesculus , Fraxinus, Salix elegans, S. australis, Ulmus wallichiana, Acer pictum, Padus cornuta.*
   - (c) 2500 m and above : pasture-lands.
4. Subalpine 2300-3300 m.

Abies pindrow, dominant at lower elevations, replaced higher up by *Abies spectabilis, P. griffithii* dominant on southern exposed sits, *Picea smithiana* rare.

Desmodium, Indigofera bushes, (*Betula*).

5. Alpine, 3300 m and above

Betula utilis, Salix kareliana, Rhododendron campanulatum.

The vegetation profiles of a few zones of Kashmir are given here (Fig.18).

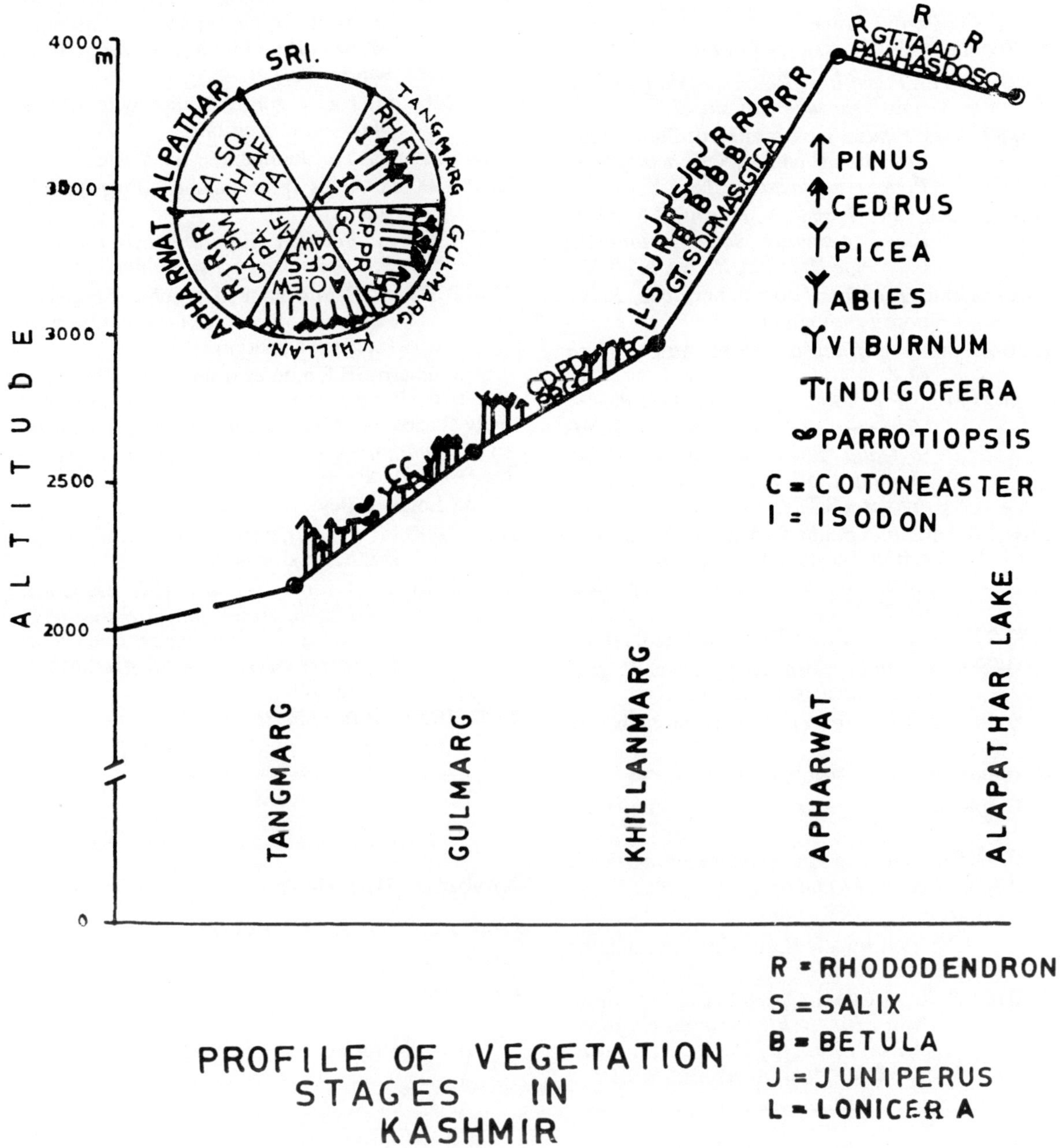

Fig. 18 : Profile of vegetation stages in Pir Panjal range Kashmir (cf Dhar ,1978).

1. Pir-Panjal range

1750-2500 m: mostly denuded forests.

2000-2150 m: *P.griffithii, Picea smithiana, Parottiopsis jacquemontiana*, *SLY**

2000-2500 m: *Pinus, Picea* (scattered) *Cotoneaster, Aesculus, Ailanthus, Salix acmophylla, Populus nigra italica*, NELY

2000-2600 m: Forest patchy, *P. griffithii, Abies, Picea, Aesculus, Salix acmophylla, Crataegus, Robinia, Padus*, NWLY,

2000-2250 m: *P.griffithii* dominant. *Picea, Parottiopsis* throughout.

2250-2500m: *P.griffithii* dominant, *Abies, Picea. ELY*

2300-2450 m: *P.griffithii, Abies* (far less dense), *Viburnum nervosum, Berberis royleana, Rosa macrophylla, Rubus niveus, Isodon plectranthus, Indigofera.* SELY

2300-2700 m: *Abies pindrow* (dominant), *Picea, P. griffithii* scattered, *Taxus baccata* isolated, *Juglans, Viburnum, Parottiopsis.* NLY

2550-2750 m: *Abies, Picea, Taxus* isolated. NLY

2600-2900 m: *Abies, Picea* insignificant, *P. griffithii* scarce, isolated. NELY.

2700-3000 m: *Abies* (pure), *Picea* incon-spicuous. NLY

2900-3400 m: Abies, Betula isolated.SELY

3300-3600 m: *Juniperus, Cotoneaster, Viburnum.* NELY

3600-3900 m: (meadows), *Cotoneaster, Gaultheria nummularioides*

2. Sind valley (Dar, 1986).

1600 m: lake, rich aquatic flora nearly choking the water body.

1850 m: *Salix (Robinia pseudoacacia), Ailanthus altissima, Rosa macrophylla, Zizyphus jujuba. (cultivated: cereals, Juglans, Aesculus, Padus).*

1850-2500 m: *Picea smithiana, Juglans, Aesculus, Padus.*

1900-2800 m: *Cedrus deodara. P.griffithii* (less frequent), *Abies pindrow, Salix* sps. *Morus alba, Juglans, Padus, Ulmus wallichina.*

2300-2600 m: rocky gorges, lithophytic *(Abies pindrow, P. griffithii).*

2100-3500 m: *Abies* dominant, *P. griffithii.*

2500-3300 m: *Abies* (pure patches) *P. griffithii* scattered

2700-2900 m: *Betula utilis, Juniperus recurva, Rhododendron hypenanthum.*

3300-3700 m: *Betula* stunted shrubs, *Juniperus recurva,* luxuriant *Rhododendron* (meadows, margs).

Southern-SE,E mountain slopes dry, "barren", only tree element poor blue pine, stunted often, many shrubs like *Plectranthus rugosus, Artemisia vestita, Indigofera heterantha, Daphne oleoides, Rosa webbiana.*

3. Lolab valley

1600-2000 m: *P. griffithii* pure patches on southeren slopes. 21%.

1840-2615 m: *Cedrus deodara* 46% associate *P.griffithii, Abies pindrow, Juglans, Aesculus indica, Acer caesium, Fraxinus exelsior, Parrotiopsis* bushes throughout.

2460-3075 m: *Abies pindrow* at lower altitudes, *P. griffithii* along ridges and depressions, *Juglans, Aesculus, Acer.*

3300 m and above: *Salix flabellaris, Betula, Juniperus.*

Kashmir -Himalaya

Genera with single species

SPECIES NOT COMMON WITH THE ADJOINING AREA

Acroglochin persicarioides (Poir)Moq
Aegilops tauschii Cossen
Aegopodium alpestre Ldb
Agroelymus koelzii Molderis
Alteris pausciflora **(Klotz) Hand-Mazz**
Alloteropsis cimicina **(L)Stapf**

* SLY southerly. NLY northerly. ELY easterly. SELY southeasterly. NELY northeasterly NWLY northwesterly.

Andropogon tristis Nees ex Hsck
Angelica glauca Edgew
Aphragmus obscurus (Dunn)OES
Arceuthobium minutisimum HK f
Archangelica officinalis Hoffm
Arctium lappa L
Arceosperma primulifolium (T T)OES
Arthrophytum thomsonii (Bge ex Boiss)iljin
Boenninghausenia albiflora (HK)Rechb ex Heynh
Bosea amherstiana HK f
Briza maxima L
Cardaria draba (L) Desv
Cardiocrinum giganteum (Wall)Makino
Carum carvi L
Cassiope fastigiata (Wall)D Don
Catapodium regidum (L) CE Hubb
Ceratoides papposa Botsch et Ikonn
Cheirostylis griffithii Lindl
Chloris dolichostachya Lagasca
Christocolea crassifolia Camb
Chrysoplenium trichospermum Edgw ex HKT
Colchicum luteum Baker
Coleus forsskohlii (Willd)Briq
Conringia planisiliqua F et M
Coriaria nepalensis Well
Cortia depressa (Don)Norman
Cortusa brotheri Pax ex Kipsky
Cryptolepis buchanarii Roem et Schult
Cypripedium cordigerum D Don
Dactylorhiza hatagirea (D Don)Soo
Desmostachya bipinnata (L) Stapf
Dichanthium annulatum (Forssk)Stapf
Dontostemon glandulosus (K et K)OES
Dregla volubilis (L F)Bth & HK f
Dubyaea oligocephala (Sch-Bip)Stebbins
Eccoilopus cotulifer (Thunb) A Camus
Epimedium elatum Morr & Dcne
Eremopoa persica (Trin) Roshev
Eulaliposis linata (Retz)CE Hubb
Eryale ferox Salisb
Filago pyramidata L
Filipendula vestita (Wall ex G Don)
Fritillaria imperialis L Maxim
Garhadiolus minutissimus (Bge)Kitam
Gastroda orobanchoides (Falc)Bth
Gaultheria trichophylla Royle
Glyceria tonglensis CB Cl
Goldbachia laevigata (MB) DC
Gymnadenia orchidis Lindl
Halenia elliptica D Don
Halocharis violacea Bge
Haloxylon thomsonii Bge ex Boiss
Hammada cashemeriana Iljin
Hemathria compressa (L f)R Br
Hierochloa laxa R Br
Holboellia latifolia Wall
Holostemma annularis (Roxb)K Schum
Hypecoum leptocaryum HK & T
Hypoxis aurea Lour
Incarvillea emodi (Lindl) Chatterji
Isachne himalaica HK f
Isatis costata CAM
Ixiolirion montanum (Lab)Herb
Koelpinia linearis Pallas
Karthalsella opuntia (Thunb)Merrill
Lagurus ovatus L
Lancea tibetica HK f & T
Lathraea squamaria L
Lignariella obscura (Dunn) Jafri
Lilium polyphyllum D Don
Linaria dalmatica (L) Miller
Liriope graminifolia (L) Baker
Lotus corniculata L
Malcomia africana (L)R Br
Marsfenia roylei Wight
Milium effusum L
Miscanthus nepalensis (Trin)Hack
Monotropa hypopitys L
Myosoton aquaticum L Moench
Natholirion thomsoninanum (Royle)Stapf
Nardurus martimus (L) Murbeck
Neyrandia arundinacea (L)Henr
Olgae thomsonii (HK f)Iljin
Onopordum acanthium L
Orinus thoraldii (Stapf) Bor
Paeonia emodi Wall ex HK f
Parrotiopsis jacquemontiana (Dcne)Rehder

Pegaeiphyton scapiflorum (HK & T) Marq et Shaw
Petasites tricholobus Franchet
Petrohagia alpina (Habl)PW Bell & Heywood
Phaeonychium parryoides (Kurz) OES
Phacelurus speciosus (Steud)CE Hubb
Phryma leptostachya L
Phytolacca latbenia (Moq)Hans
Phyla nodiflora (L) Greene
Physochlaena prealta (D Don) HK F
Picris hieraciodes L
Picrorhiza kurroa Royle ex Bth
Pleconema griffithiana (Boiss) Roch
Plumbago rosea L
Podophyllum emodi Wall ex Royle
Polemonium coerulum L
Portulaca oleracea L
Pseudostellaria himalaica (Franch) pax
Pycnoplinthus uniflorus (HK & T)OES
Pulsatilla wallichiana (Royle) Ulbr
Rhynchospermum verticillatum Reinw
Sambucus wightiana Wallex Wight & Arnott
Sanicula elata Ham
Sarcicca saligna (Don) Muell
Satyrium nepalense D Don
Schizachyrium impressum (Hack) A Camus
Scleropoa rigida (L) Griseb
Scoparia dulcia L
Serratula pallida DC
Seseli libanotis (L) Koch
Sium latijugum Cl
Skimmia lareola (DC) Sieb et Zucc ex Walp
Solena heterophylla Lour
Solidago vergaurea L
Sophora alopecuroides L
Soroseris deasyi (S Moore)Stebbins
Spiranthes lancea (Thunb)Baker
Stracheya ţibetica Bth
Tetracme pamirica Vass
Thymus serpyllum L
Torularia humilis (CAM)OES
Trigonotis tibetica (CAM)M Johns
Trillium govanianum Wall ex Royle
Trikeraia hookeri (Stapf)Bor
Viscum album L
Vulpia myurus (L)Gmel
Wilkstroemia canascens Meissn
Woodfordia fruticosa (L)Kurz
Wulfenia amherstiana Wall ex Bth
Xylanche himalaica (HK & T)G Beck

ADDITIONAL SPECIES COMMON TO KASHMIR AND KAZAKHSTAN

Adoxa moschatellina L
Berula erecta (Huds)Cavill
Buchingerca axillaris Boiss
Caltha palustris L
Cichorium intybus L
Cicuta virosa L
Cimicifuga foetida L
Corallorhiza trifida Chatel
Eruca sativa Mill
Goodyera repoens (L)R Br
Leptorhabdos parviflora (Bth)Bth
Llyodia serotina (L)Rchb
Menyanthes trifoliata L
Monesis uniflora (L)A Gray
Siegesbackia orientalis L
Silybum marianum Gaertn(L)
Thalacospermum caespitossum (Camb) Schischk
Thymaelea passerina (L) Coss et Germ
Trachyspermum ammi (L)Sprangne
Tussilago farfara L
Vacaria pyramidata Medic

ADDITIONAL SPECIES COMMON TO KASHMIR AND UZBEKISTAN

Atelanthera purpussilla HK & T
Callitriche palustris L
Dopatrium junceum (Roxb)Ham ex Bth
Eruca sativa Miller
Halogeton glomerata (MB)CAM
Lapsana communis L
Listera ovata (L)R Br
Oxalis corniculata L
Samolus valearndi L
Silybrum marianum (L)Gaertn
Sideritis montana L
Tauscheria lasiocarpa Fish ex DC

Trachyspermum ammi (L)Sprague

ADDITIONAL SPECIES COMMON TO KASHMIR AND KIRGIZ

Cicuta virosa L
Cichorium intybus L
Coeloglossum viride (L)Hartm
Dilophia salsa T T
Erophila verna (L)Besser
Hedina tibetica (Thoms)St Ostenf
Limosella aquatica (L. (+ KZ)
Leersia oryzoides (L)Sw
Prunella vulgaris L
Sideritis montana L
Siegesbeckia orientalis
Thalacospermum caespitosum
Tribulus terrestris L (+ Tdj)
Thymaelaea passerina (L)Coss et Gaertn

ADDITIONAL SPECIES COMMON TO KASHMIR AND TADJIKSTAN

Atelanthera purpusilla HK f & T
Buchingera axillaris Boiss
Carum carvi L
Cephalanthera longifolia (L)Fritsch
Ceratophyllum demersum L
Crypsis schenoides (L)Lamk
Dilophia salsa Them
Duchesnia indica (Andr)Focke
Eurica sativa Miller
Halogeton glomeratus (MB)CAM
Hedinia tibetica (Thoms)Ostenf
Leersia oryzoides (L)Sw
Lloydia serotina (L)Rehb (+ Kg)
Samolus valerandi L
Tauscheria lasiocarpa F ex DC
Thymalaea passerina Coss et Germ

4

ENDEMISM

CENTRAL ASIA

Ikonnikov (1965) studied the flora of Pamir and stated that 4 percent of plant species are endemic in this region! and that these endemics are of recent origin, belonging mostly to the polymorphic families: Compositae, Chenopodiaceae, Boraginaceae, Gramineae, etc. Some examples of endemic species in Pamir are : *Taraxacum pamiricum, Tragopogon pamiricum, Eritrichum dubium, E. pseudostrictum, Chenopodium pamiricum, Agrostis pamirica.* Further a comparative study on plant distribution in the Himalaya, Tibet and Pamir has revealed that the flora is improverished from south to north in the number of families, genera and species; perhaps owing to the severe conditions of climate prevailing in Pamir.

Pavlov (1970) recorded 13% endemism in the total flora of the Tien Shan, the majority being restricted to definite types of habitats, about two-third of all the endemics inhabit the sites which are in the initial stages of rock degredation.

In the high altitude areas of Caucasus and Central Asia the endemism is very feeble (Avetisyan, 1974). Only one genus is endemic in Caucasus while in Central Asia just two. The specific endemism is rich: 54% in the Caucasus and 60% in Central Asia; further that the remarkable specific endemism has been provided by the lofty mountain chains.

In Iran Zohary (1963) stated that ca 20% of the total number of species are endemic and these are "progressive and very young" because Iran is one of the centres of origin of Irano-Turanian flora; though a secondary one and a vast number of endemics occur as groups, allied species clusters, the distribution of which being either allopatric or sympatric; in the former each member of cluster occupies a different habitat while reverse holds true for the latter.

The examples of both of these types are *Astragalus, Acantholimon, Heliotropium, Cousinia,* Chenopodiaceae, Cruciferae, Leguminosae, Umbelliferae, Boraginaceae, Labiatae, etc, the characteristic endemic taxa of the area. The genera rich in endemics are *Silene, Canthophyllum, Delphinium, Erysimum, Isatis, Amygdalus.*

It may be pertinent to refer to the endemism in Alps and Carpathian mountains which Pavlowski (1970) found rich in Palaeo- and Neo- endemics and that European mountains in general. are richest in endemics (esp ancient endemics). Here Favager (1972) found that the southern mountains are richer in endemics (30-40 %) while the northern ones including Pyrennes, eastern and western Alps, Carpathian and even central Appennines have a low endemic percentage (12-18). The Alps represent two centres of endemism, in the northeast southwest Alps and in southeast and the northeast calcareous Alps both have high percentage of endemism (30,46 % respectively). Contandriopoulous (1962) stated that (a) there exists a north to south gradient of endemism, probably due to more favourable climatic conditions in the Mediterranean region. (b) the mountains of southern Europe are believed to have been slightly affected by the glaciation and could have provided refuge for those species which became extinct elsewhere, and (c) the central European mountains have a low proportion of endemism, owing to a higher proportion of mountain species in common with other adjacent mountain chains.

Western Tien Shan (Pavlov, 1950): The western Tien Shan is one of the richest part of Middle Asia and the number of endemic plant species is about 365, which is 13% of the total numberof the species of the flora of this region (Table I). The majority of " the endemics are restricted to definite types of habitats, thus ca two-thirds of all the endemics inhabit areas characterised by the initial and/or medium stages of rock destruction". It is concluded that the "present day dominance of neo-endemics from the polymorphic genera of the ancient Mediterranean and Middle Asian origin is shown with a certain participation of ancient mesophilic relict endemics gravitating to the relic communities. Such a character of endemism is the result of the effect of the orogenesis and of the aridization of the climate on the vegetational cover and of the heterogeneity of the flora".

Table I: Endemism in flora of western Tien Shan (endemic species/total species).

Pinaceae: *Abies* 1/1

Cupressaceae: *Juniperus* 3/8

Gramineae: *Stipa* 2/16, *Pentatherum* 1/3, Agrostis 1/6, *Poa* 1/21, *Eremopoa* 1/4, *Bromus* 2/18, *Agropyron* 1/21, *Elymus* 3/9.

Liliaceae: *Eremurus* 3/9, *Gagea* 3/21, *Allium* 11/54, *Tulipa* 11/19

Iridaceae: *Juno* 3/7

Salicaceae: *Populus* 1/16

Betulaceae: *Betula* 1/8

Urticaceae: *Parietaria* 1/4

Santalaceae: *Thesium* 2/3

Polygonaceae: *Rumex* 1/15, *Calligonum* 1/1, *Polygonum* 4/36

Caryophyllaceae: *Arenaria* 1/6, *Silene* 5/26, *Dianthus* 4/16

Ranunculaceae: *Delphinium* 1/14, *Aconitum* 1/4, *Ranunculus* 3/34, *Adonis* 1/3.

Cruciferae: *Erysimum* 5/14, *Arabis* 3/10, *Pseudoclausia* 2/6, *Parrya* 7/17, *Christolea* 1/1, *Draba* 1/19, *Crambe* 1/3, *Stroganovia* 1/3, *Stubendorffia* 1/4.

Crassulariaceae: *Rhodiola* 1/6, *Rosularia* 1/6.

Saxifragaceae: *Bergenia* 1/1.

Rosaceae: *Exochorda* 1/1, *Malus* 4/7, *Crataegus* 1/7, *Rubus* 1/4, *Sanguisorba* 1/3, *Rosa* 1/20, *Prunus* 1/2, *Amygdalus* 2/4, *Cerasus* 1/5.

Papilionaceae: *Medicago* 1/10, *Calophaca* 1/1, Chesneya 1/2, *Astragalus* 36/121, *Tragacanthus* 2/6, *Oxytropis* 21/48, *Glycyrrhiza* 1/4, *Cicer* 1/4, *Hedysarum* 13/22.

Euphoriabiaceae: *Euphorbia* 3/24

Vitaceae: *Vitis* 1/2.

Malvaceae: *Alcea* 1/4

Violaceae: *Viola* 1/12

Umbelliferae: *Schrenkia* 3/6, *Prangos* 1/5, *Bupleurum* 1/6, *Ligusticum* 1/2, *Selinum* 4/4, *Bunium* 1/5, *Scaligera* 4/10, *Libanotis* 5/8, *Sphenocarpus* 1/1 *Aulacospermum* 1/5, *Trachydium* 2/2, *Peucedanum* 1/1, *Talassia* 1/2, *Ferula* 9/26, *Dorema* 1/1

Primulaceae: *Androsace* 1/9

Plumbaginaceae: *Acantholimon* 5/12

Gentianaceae: *Swertia* 1/5

Convolvulaceae: *Convolvulus* 1/8

Boraginaceae: *Heliotropium* 1/4, *Onosma* 1/4, *Lepechiniella* 1/8, *Stephanocaryum* 1/2, *Rindera* 5/11, *Trachelanthus* 1/1, *Cynoglossum* 1/3.

Labiatae: *Scutellaria* 6/17, *Lophanthus* 1/2, *Nepeta* 2/13, *Pseudoeremostachys* 1/1, *Dracocephalum* 3/16, *Eremostachya* 6/13, *Phlomis* 6/10, *Lagochilus* 3/8, *Stachys* 1/5, *Salvia* 2/7, *Schradera* 1/1, *Thymus* 1/8.

Scrophulariaceae: *Linaria* 1/4, *Scrophularia* 2/11, *Euphrasia* 1/5, *Pedicularis* 4/14. Rubiaceae: *Asperula* 1/3, *Rubia* 1/3.

Caprifoliaceae: *Lonicera* 1/13.

Campanulaceae: *Asyneuma* 1/3, *Sergia* 1/1.

Compositae: *Ugamia* 1/1, *Trichanthemis* 4/7, *Lepidolophia* 2/3, *Tanacetopsis* 2/4, *Senecio* 2/8, *Ligularia* 3/10, *Echinops* 1/8, *Cousinia* 29/52, *Hypecanthium* 1/1, *Saussurea* 4/9, *Jurinea* 7/14, *Olgea* 1/1, *Serratula* 1/5, Syreitschikovia 1/1, *Rhaponticum* 2/3, *Centaurea* 2/3 3/10, *Scorzonera* 1/13, *Taraxacum* 1/29, *Lactuca* 1/6, *Cephalorrhynchus* 1/2, *Cicerbita* 2/4.

Middle Asian Mountain Province

Kamelin (1965) studied generic endemism in Middle Asian flora and concluded that it deserved to be recognised as a separate province which he named as Middle Asian Mountain Province. He also outlined its boundaries. His list of such genera is given below (all monogeneric, unless where indicated) .

Liliaceae: *Korolkowia* Rgl.

Chenopodiaceae: *Rhaphydophyton* Iljin, *Physandra* Botsch, *Arthrophytum* Schrenk (8).

Fumariaceae: *Fumariola* Korsch.

Cruciferae: *Catenularia* Botsch, *Sisymbriopsis* Botsch, *Iskandera Busch, Trichochiton* Kom, *Sprygima* M Pop.

Rosaceae : *Spiraeanthus* Maxim, *Aflatunia* Vved.

Umbelliferae: *Albertia* Rgl et Schmath, *Scheschrovskia* Rgl et Schmath, *Kosopoljanskia* Korov, *Lipskya* (K-Pol) Nevski, Sclerotiaria Korov, *Astomatopsis* Korov, *Schrenkia* Fisch et Mey(10-11), *Buniella* Schischsk(2), *Zeravschanica* Korop, *Hymneolyma* Korov (2), *Korshinskya* Lipsky (2), *Ladyginia* Lipsky, *Komarovia* Korov, *Hyalolaena* Bge (14), Mogoltavia Korov, *Scaphospermum* Korov, *Pilopleura* Schischsk, *Vvedenskya* Korov, *Paulia* Korov, *Talassia* Korov (2).

Plumbaginaceae: *Chaetolimon* (Bge)Lincz (3-4).

Boraginaceae: *Ulugbekia* Zak, *Stephanocaryum* M Pop.

Labiatae: *Apellanthus* Nevski ex Pavl (10), Kudrjaschevia Pojark (3), *Pseudoeremistachys* M Pop, *Pseudomarrubium* M Pop.

Scrophulariaceae: *Spirostegia* Ivanina, *Nathaliella* B Fedtsch.

Bingnoniaceae: *Niedzwedzkia* B Fedtsch.

Campanulaceae: *Ostrowskia* Rgl, *Sergia* Fed (2), *Croyptocodon* Fed., *Cylindocarpa* Rgl.

Compositae: *Trichanthemis* Rgl et Schmath (5), *Lapidolphia* C Winkl (4), *Cancriniella* Tavel, *Modesta* Tamamsch et Charadse (2) *Polychrysum* S Koval, *Ugamia* Pavl, *Polytaxis* Bge (2), *Anura* (Juz)Tschen, *Lamyropappus* Knorr, *Hypacanthium* Juz, *Schmathausenia* CWinkl.

AFGHANISTAN

Wandelbo (1966) recorded the following position in endemism (in percentages) in Afghan flora (total species withim brackets):

A. FAMILIES

(53) Plumbaginaceae 75
(400) Compositae 40
(300) Leguminosae 40
(160) Labiatae 35
(34) Primulaceae 35
(170) Cruciferae 20
(110) Liliaceae 18
(15) Campanulaceae 33
(140) Umbelliferae 30
(120) Boraginaceae 30
(110) Caryophyllaceae 30
(75) Scrophulariaceae 30
(40) Rubiaceae 15
(14) Onagraceae 7

B. GENERA

Dionysia 90
Acanthophyllum 83
Cousinia 80
Acantholimon 75
Mattiastrum 70
Astragalus 70
Scrophularia 53
Ferula 45

HINDÜKUSH

Breckle(1974) stated that ca 20.02% of the flora of Hindukush is endemic (based on species flourishing at or above 4000 m). Further that the number of endemics declines with the increase in elevation, probably due to the presence of similar environmental conditions prevailing on all high mountain chains, making them widespread in distribution. He recorded 82 endemic species in 20 families in the alpine Hindukush region. Leguminosae alone having 18 endemic species followed by Compositae, Boraginaceae, etc. According to Kitamura (1955) Afghanistan became arid rather recently with the result no hygrophyllous plants could exist there. Thus many young endemics were evolved from the most changeable groups that had invaded from the neighbouring lands without competing with the old endemics in this area. Breckle (1974) listed the following endemics in alpine Hindukush region including Chitral and Afghan-Wakhan.

Umbelliferae: *Platytaenia lasiocarpa* (Bss) Rech f et Riedl ssp *lasiocarpa* and ssp *incana* Rech f et Reidl.

Compositae: *Cousinia riana* Bornm, *C. blepharobasis* Rech f et Gilli, *C. chionophila* Rech f et Koeie, *C. carthamnoides* Aitch et Hemsl, *C. kotandarica* Rech f et Gilli, C. polyneura Rech f, Bornm, *Erigeron petroiketes* Rech f, *Saussurea chthonocephala* Bornm, *S.karstanii* (Bornm) Rech f.

Boraginaceae: *Arnebia speciosa* Aitch et Hemsl, *Eritrichium afghanicaum* Rech f, *A. rechingheri* Riedl, *Lepichinella albiflora* Reidl *Mathiastrum deilsii* Bornm, *Omphlodes heterophylla* Rech f et Reidl, *Pseudomertensia lindelofioides* (Rech f et Riedl) Riedl, *Solenanthus strictiissimus* Brand, *S. micranthus* Riedl, *Tianshaniella wakhanica* Riedl

Cruciferae: *Chorispora pectinata Hadec, Draba afghanica Bss, Erisymum griffithii (Hk f et T) Jafri, Fibigia membranacea Rech f, Grellsia chitralensis OES, Gynophorea weileri Gilli, Stragonowia puberula Kitam.*

Caryophyllaceae: *Silene daniellii* ~~He~~dac.

Cyperaceae: Carex chitralensis Nelmis, C. griffithii Boott.

Ericaeae: *Rhododendron collecttianum* Aitch et Hamsl.

Papilionaceae: *Astragalus ajfreidii* Aitch et Baker, ssp *ajfredii*, ssp *breviovexillatus* Deml, *A. aloisii* Deml. *A. confertissimus* Kitam, *A. gregarius* Deml, *A. hassarorum* Gilli, *A. hindukushensis* Wdb. *A. macrostegius* Rech f, *A. minutifoliolatus* Wdb, *A. miseriflorus* Sirj et Rech f, *A. peltatus* Podi et Deml, *A. rassoulii* Riedl, *Cicer nuristanicum* Kitam, *C. pungens* Bss, *Hedysarum minjanensis* Rech f, *Oxytropis afghanica* Rech f et Koeie, *Trigonella afghanica* Vass.

Fumariaceae: *Corydalis fedtschenkoana* Schrenk ssp *metallica* (Wdb) Wdb

Gentianaceae: *Gentiana micrantha* Aitch et Hemsl.

Labiatae: *Dracocephalum glecomifolium* Dunn. *Eremostachys colophyta* Hdg. *Thymus afghnica* Ronn.

Juncaceae: *Juncus triglumis* L. ssp *wakhaniensis* Snog.

Liliaceae: *Gagea setifolia* Baker, *Allium nuristanicum* Kitam.

Plumbaginaceae: *Acantholimon auganum* Bge, *A. calocephalum* Aitch et Hemsl, *A. diapensioides* Bss, *A. hariabense* Rech f et Koeie, *A. peculiare* Rech f.

Gramineae: *Agrostis interjacens* Melderis, *Colopodium afghanicum* Bor, *Festuca afghanica* Bor, *Poa roemeri* Bor, *Psatyrostachys caduca* (Bss) Melders.

Polygonaceae: *Polygonum chitralicum* Rech f et Sch-Cz.

Ranunculaceae: *Ranunculus shaftoanus* (Aitch & Hemsl) Bss, *Thalictrum afghanicum* Gilli, *Th.parvulum* Rech f

Rosaceae: *Potentilla coelestis* Gilli, *P. collectiana* Aitch & Hemsl. *P. kurramensis* Th Wolf.

Scrophulariaceae : *Pedicularis afghanica* Wdb.

THE HIMALAYA

Chatterjee (1939) recorded 3,165; 2,045 and 1,071 endemic species in the Himalaya, continental undivided India, and Burma respecstively; and that the high endemic figures of these three regions were suggestive of their young age being produced from a stock capable of change. Besides such taxa as never found an opportunity or enough time for their migration and differentiation; consequently remained localised in distribution. An approximate analysis of the Indian dicots revealed 61.05 % of endemics and only 38.05 % of the total plants as wides. He divided Indian dicots (173 families) into the following groups:

(a) families with less than 20 species in each, mostly widespread. 80

(b) families with 20 or more species in each and with a majority of more than 50 % species non-endemic or wides 27

(c) families with 20 or more species in each and with a majority of more than 50% of the species endemic . . . . .65.

He provided the following figures in endemism of some major families of the region (percentage given)

Compositae 52
Fumariaceae 3 70
Balsaminaceae 91
Rosaceae 70
Rubiaceae 67
Labiatae 62
Cruciferae 52
Papaverceae 96
Papilionaceac 57
Saxifragaceae 76
Ericaceae 90

Meher-Homji and Misra (1973) also found India rich in endemic species, but not so in endemic genera (7.3 % only). The specific endemism given was 20.8%, 18.2 % and 9.6 % in the Himalaya, peninsular India, and Burma, respectively. Only 4.9 % of the species have been found as endemic to the whole of the subcontinent.

Takhtajan (1969) listed ca 150 endemic genera in the India floristic region (which includes peninsular India, Himalayan flanks and Sri lanka); Rao (1972) recognised 164 endemic genera in the same region (including Burma) and Nayar (1980) recorded 141 endemic genera exclusively within the boundry of present India, in which 111 genera are monotypic.

"Besides there are ca 109 endemic genera which have their distribution in the Himalaya, Himalayan-Tibet, SW China, NE India-Burma-Yunan triangle" (Nayar 1980); of these he listed 68 in the Himalaya (as given below, the number of species where more than 1 is appended):

WESTERN HIMALAYA: *Arcyosperma* Schulz (Cruciferae), *Catamixis* Thoms (Compositae), *Falconeria* Hook f (Scrophulariaceae), *Ischnochloa* Hook f (Gramineae), *Meeboldia* H. Walf(Umbelliferae), *Microschoenus* CBCl (Cyperaceae), *Parrotiopsis* (Neid)Schmeid (Hamamelidaceae) *Picrorhiza* Royl ex Bth (2 Scrophulariaceae) *Platystemma* Wall(Gramineae), *Pseudodanthonia* Bor & CE Hubbard(Gramineae), *Sempervivella* Dunn (4,Crassulaceae).

EASTERN HIMALAYA: *Biswarea* Cogn (Cucurbitaceae), *Brachystemma* D.Don (Caryophyllaceae + Central Him), *Bryocarpum* Hook f & T (Primulaceae), *Cautleya* Royle (5,Zingiberaceae), *Cavea* WW Smith & Small (Compositae), *Cleisocentrom* Bruhl (Orchidaceae + Assam), *Cucumorpha* AS Rao & DM Verma (Zingiberaceae + Assam), *Cyathopus* Stapf (Gramineae), *Drimycarpus* Hook f (Anacardiaceae), *Edgaria* CB Cl (Cucurbitaceae), *Esomesdraldia* Reich f (3,Orchidaceae), *Lacaitaea* Brand (Boraginaceae), *Lepidostemon* Hook f & T (Cruciferae), *Listrobanthus* Bremek (Acanthaceae + Assam), *Leptocardon* Lem (Campanulaceae), *Milula* Prain (Alliaceae), *Neoluffa* Chakrav (Cucurbitaceae), *Parajaeschkea* Burkill (Gentianaceae), *Poroxygraphis* WW Smith (Ranunculaceae), *Pauia* Deb et Dutta (Solanaceae), *Pedinogyne* Brand (Boraginaceae), *Pleurospermopsis* C Norman (Umbelliferae), *Pollysolenia* Hook f (Rubiaceae), *Pseudostachyum* Munro (Gramineae), *Smithiella* Dunn (Urticaceae), *Stilbanthus* Hook f (Amaranthaceae), *Stiptanthus* (Bth)Briq (Labiatae), *Sympagis* (Nees) Bremek (5,Acanthaceae), *Tarphochlamys* Bremek (Acanthaceae + Assam), *Treutlera* Hook f (Asclepiadaceae), *Triaenanthus* Nees (Acanthaceae + Assam).

HIMALAYA: *Cortia* Hook f (Umbelliferae), *Didiciea* King & Prain (Orchidaceae), *Jaeschkea* Kurz (3, Gentianaceae), *Megacodon* (Hemsl)H Smith (Gentianaceae), *Notochaete* Bth (Labiatae), *Pseudaechmanthera* Bremek (Acanthaceae), *Pteracanthus* (Nees)Bremek (20, Acanthaceae), *Pycnoplinthus* OES (Cruciferae), *Roylea* Wall (Labiatae), *Thomsonia* Wall (Araceae).

ALPINE HIMALAYA: *Chionocharis* IM Johnson (Boraginaceae), *Eriophyton* Bth (Labiatae).

TEMPERATE HIMALAYA: *Aechmanthera* Nees (3, Acanthaceae), *Craniotome* Reichb (Labiatae), *Crytochilus* Wall (2, Orchidaceae), *Hemiphragma* Wall (Scrophulariaceae + Khasia Hills), *Theropogon* Maxim (Liliaceae).

TROPICAL HIMALAYA: *Diplomeris* D Don (3, Orchidaceae).

ASSAM: *Indofevillea* Chatterjee (Cucurbitaceae), *Khasiaclunea* Ridsd (Rubiaceae + Khasia), *Micholitzia* NE Br (Asclepiadaceae), *Parakaemferia* A S Rao & D M Verma (Zingiberceae), *Pentabothra* Hook f (Asclepiadaceae), *Polyura* Hook f (Rubiaceae + E Him).

KHASIA & JAINTIA HILLS: *Caulokaempferia* Kai Larsen (2, Zingiberaceae)

NAGALAND: *Pottingeria* Prain (Escalloniaceae).

Kashmir

Dhar (1978) recorded 589 species in the alpine-subalpine Kashmir Himalaya as endemics: commenting that in order to assess the extent of endemism in a particular family it is plausible to take both the species and the varieties, putting them together and the species alone into consideration.This is illustrated in Table VIII. The extent of endemism is almost negligible in the monocots (Gramineae excluded), on the average being 15.94%. The endemic taxa in dicots are 31.36% of the total flora of Kashmir Himalaya.

Interestingly most of the alpines of this area are invariably endemics. The relationship between these two aspects in some prominent families is given in Table VIII. Dhar and Kachroo (1983b) suggested that 40% of these endemic and other restricted species in Kashmir Himalaya are threatened (Figs 19-23) owing to (a) the unhindered commercial exploitation of medical and ornamental plants (some have fodder value) and (b) biotic changes in ecology of the habitat. The number of species endemic in various genera in flora of Kashmir is given in Table II.

Table II. Range of endemism in the flora of Kashmir Himalaya (based on Dhar and Kachroo, 1983 a) (total species/endemic species).

Ranunculaceae: *Aconitum 3/1, Anemone 10/1, Aquilegia 5/2, Delphinium 6/1+3***, Paraequilegia 2/1***, Ranunculus 27/3+4***, Trollius 1/1**.

Berberidaceae: *Berberis 16/3.*

Papaveraceae *Meconopsis 1/1*$^{+}$.

Fumariaceae: *Corydalis 25/ 4+7**,

Cruciferae: *Aphragmus 1/1*$^{+}$*, Arabidopsis 5/1*$^{+}$*, Arabis 9/1+1***, Braya 5/1, Dotnostemon 1/1, Draba 18/ 3+2***, Ermania 5/3+1***, Lignariella 1/1***, Megacarpae 2/2, Pycnoplinthus 1/1, Thlaspi 5/1**.

Violaceae: *Viola 12/ 2+ 4**.

Caryophyllaceae: *Arenaria 10/ 2+1***, Dianthus 7/2, Lychnis 11/ 1+1***, Stellaria 12/ 2+1**.

Geraniaceae: *Geranium 18/ 2**

Balsaminaceae: *Impatiens 14/ 1+3***+2*$^{+}$

Papilionaceae: *Astragalus 47/25 Caragana 5/1. Chesneya 3/1***, Colutea 1/1***, Hedysarum 6/1 +3***, Oxytropis 14/5***+1*$^{+}$*, Stracheya 1/1, Thermopsis 2/1. Trigonella 7/2*$^{+}$.

Rosaceae: *Cotoneaster 11/2***, Potentilla 47/10+3**.

Crassulaceae: *Sedum 19/ 3+1***, Sempervivella 3/1**.

Saxifragaceae: *Chrysosplenium 1/1***, Saxifraga 25/ 4***+ 3*$^{+}$.

Onagraceae: *Epilobium 21/ 3 +4*$^{+}$.

Umbelliferae: *Angelica 1/1*$^{+}$*, Anthriscus 2/1.Bupleurum 17/2+ 2***+2*$^{+}$*, Chaerophyllum 5/1+1*$^{+}$*, Cortia 1/1, Eriocycla1/1***,Heracleum 5/1***+1*$^{+}$*, Peucedanum 1/1 , Pleurospermum 7/ 3** *+ 3*$^{+}$*, Schulzia 1/1***, Selinum 6/1+1** *+2*$^{+}$*, Trachydium 1/1*$^{+}$*, Vicatia 2/1*$^{+}$.

Caprifoliaceae: *Lonicera 18/3+1** *,+1*$^{+}$*, Sambucus 1/1***, Viburnum 3/1**

Valerianaceae: *Valeriana 8/4**.

Rubiaceae: *Galium 14/4**.

Dipsacaceae: *Dipsacus 2/1*$^{+}$*, Morina 3/1*$^{+}$*, Scabiosa 1/1**.

Compositae: *Aster 12/ 1+1***+3*$^{+}$*, Brachyactis 3/1***, Erigeron 13/1***, Psychrogeton 3/1*$^{+}$*, Anaphalis 9/1+1***, Gnaphalium 5/1, Inula 12/3***, Leontopodium 6/1+1***,Chrysanthemum 6/2***, Tanacetum 9/5***+2*$^{+}$*, Waldheimia 5/1+1*$^{+}$*, Cremanthodium 4/1***, Doronicum 4/1** *+1*$^{+}$*, Senecio 13/4***, Cirsium 7/1***, Cousinia 5/1, Jurinea 4/3***, Saussurea 31/ 6** *+ 3*$^{+}$*, Tricholepis 4/ 1+1***, Crepis 8/1*$^{+}$*, Dubyaea 1/1***, Heteropappus 1/1, Hieraceum 6/1** *Lactuca 16/4***, Scorzonera 3/1*$^{+}$*, Taraxacum 59/ 34+ 12** *+ 3*$^{+}$*, Tragopogon 4/1.*

Campanulaceae: *Campanula 8/1** + 2+, *Codonopsis 3/ 1** + 1+.

Ericaeae: *Rhododendron 3/1+.*

Primulaceae: *Androsace 14/ 1+2* +2+, Lysimachia 6/1+. Primula 21/3+4* + 4+.*

Asclepiadaceae: *Cynanchium 7/2*.*

Gentianaceae: *Gentiana 31/ 5 + 3* + 5+, Gentianella 1/1, Jaeschkea 2/2*, Lomantogonium 4/1, Swertia 13/ 1+ 5* + 2+.*

Boraginaceae: *Actinocarya 2/1, Arnebia 4/1*, Cynoglossum 7/2+ 2+, Eritrichum 5/1, Hackelia 6/ 1+2* +1+, Lasiocaryum 2/1, Lindolofia 5/3+, Mettiastrum 3/1, Microula 1/1, Pseudomertensia 8/3*.*

Solanaceae: *Physochlaena 1/1*

Scrophulariaceae: *Euphrasia 12/ 7+1*, Lagotis 3/2*, Pedicularis 27/7+ 6*, Picrorhiza 1/1+, Scrophularia 8/ 1+1* +1+, Veronica 26/ 4+4* + 1+, Wulfenia 1/1+.*

Orobanchiaceae: *Orobanche 7/1+ +1.*

Acanthaceae: *Strobilanthes 4/1*.*

Plantaginaceae: *Plantago 7/1+.*

Labiatae: *Elscholtzia 6/ 1+1+, Nepeta 28/ 6+3* + 3+, Phlomis 3/3*, Salvia 8/1+1* +1+.*

Chenopodiaceae: *Atriplex 5/1+.*

Polygonaceae: *Polygonum 41/ 5+ 2* + 1+.*

Euphorbiaceae: *Euphorbia 14/2 +1* +1+.*

Orchidaceae: *Epipactis 2/1*, Neottia 3/1+2*, Herminium 4/1+.*

Liliaceae: *Fritillaria 2/1*, Allium 19/2 + 4*, Gagea 2 4/1+.*

Araceae: *Arisaema 4/1+.*

Iridaceae: *Iris 9/2**

Juncaceae: *Juncus 12/1+.*

Cyperaceae: *Carex 61/ 5+4* + 2+, Eleocharis 7 2*, Kobresia 16/1+1* +3+.*

Gramineae: *not available.*

* Kashmir and Pakistan.

\+ Kashmir and west Himalaya.

Fig 19. *Lavatera kashmiriana*, camb., the only species in Kashmir, also occurs in Pamir.

Fig 20. *Ferula jaeschkeana* Vatke, widespread, the other species in Kashmir (*F. kokanica* Regel et Schm.) is endemic to Gilgit and South of Hindukush.

Fig 21. *Inula royleana* DC. Of the 12 species in Kashmir, *I. royleana*, *I. clarkei* (Hook. f.) R.R. Stewart and *I. acuminata* Royle ex DC. are endemic to Kashmir.

Fig 22. *Geranium wallichianum* Sw. (courtesy Bot. Surv. India)

Fig 23. *Meconopsis aculeata* Royle (courtesy Bot. Surv. India)

5

DISTRIBUTION PATTERN OF ENDEMICS IN KASHMIR HIMALAYA

FAMILIES WITH ENDEMICS

DICOTYLEDONS

Salicaceae is abundant mostly in north temperate regions, but *Salix* also extends into alpine and arctic zones. In India 43 species in two genera occur with high endemism (65.11 %) in the Himalaya (Chatterjee, 1939) ; while in Kashmir Himalaya only 22 taxa occur and in *Salix* (endemism 5 %) only two species show broad- range endemism (10.52 %). In some adjoining regions endemism is 4.16% in Pakistan, and 17.64% in Europe. Good (1964) recorded three species, namely, *Salix arctica, S. nummularia,* and *S. polaris* as endemic to the arctic and subarctic regions.

Polygonaceae, cosmopolitan, is largely temperate and comprises 100 species in eight genera in India with 77.27% endemism (Chatterjee l.c.) *Polygonum* with 41 species and *Rumex* with 11 taxa have three endemics (5%) and taxa further (21.05%) exhibit following restricted distribution .

Kashmir Himlalaya with	*No.of taxa*
Western Himalaya	2
Western and Central Himalaya	8
Eastern Himalaya	2
Total	12

In general 15 taxa (25%) reveal varying degrees of range restriction within the Himalaya and the rest (75%) extend as wides beyond this limit. In *Polgonum* alone,Chatterjee recorded 88% endemism and further opined that endemic are mostly encountered in those genera which inhabit the hilly areas.

Polygonum has endemics both in Kashmir Himalaya (7.31%) as well as Europe (23.15%); *Rumex* only in Europe (6.06%); while *Rheum* in the Eastern Himalaya (100%) and Europe (50%).

Another cosmopolitan family- **Chenopodiaceae** is represented mostly by halophytes. In India, 40 species in 18 genera with 7.5% endemism are known (Chatterjee, 1939),and 38 taxa in 15 genera occur in the study area: *Chenopodium* 10 taxa *Atriplex* with five; and eight are monotypic : *Acroglochin, Arthrophytum, Bassia, Hammada, Halocharis, Halogeton, Haloxylon* and *Ceratoides.* The endemic percentage is very low in Kashmir Himalaya (2.63) and same holds true for broad range endemics (8.10%). In general only four taxa (10.52%) exhibit restricted distribution within the Himalaya with a good number of wides. Endemism in genera of Kashmir Himalaya and the adjoining areas is presented below :

Corispermum alone reveals endemism in Kashmir Himalaya (33.33%) while *Haloxylon, Hammada* and *Arthrophytum* are each characterised by 100% broad range endemism ; and the remaining 11 genera do not display any endemism. In the adjacent areas, endemics occur as follows (percentages given) :

Pakistan: *Halocharis 20.*

Afghanistan: *Gamanthus 50, Halarchon 10, Halocharis 12.50, Kirilowia 100, Noaea 66.66.*

Soviet Asia: *Polycnemum 20, Chenopodium 100, Monolepis 100, Atriplex 24, Axyris 40, Kalidium 33.33, Agriophyllum 40, Corispermum 61.76, Halopeplis 100, Saueda 48.30, Alexandra 100, Broszezowia 100, Salsola 44.44, Rhaphidophyton 100, Horaninowia*

25, Ofaiston 100, Girgansohnia 66.66, Anabasis 47.83, Arthrophyton 71.42 Petrasimonia 54.54, Halocharis 66.66, Halimocenemis 72.72, Piptoptera 100, Halanthium 25, Gamanthus 50, Cornulaca 100.

Europe: *Polycnemum 25, Beta 14.28, Camphorosma 25, Kochia 25, Corispermum 8.33, Salicornea 55.5, Salsola 10.71.*

The overall endemic percentages for various areas are the following :

Kashmir Himalaya 2.63
Pakistan 1.40
Afghanistan 7.46
E.Himalaya nil
Soviet Asia 37.64
Europe 7.64

Predominantly a cosmopolitan family , **Caryophyllaceae** is heavily massed in the Mediterranean region . Chatterjee (l.c.) recorded 107 species in 18 genera in India with 58.57% endemism. The family is more extensive in the Himalayan region as revealed by 58.87% endemism. In Kashmir Himalaya of the 78 taxa in 21 genera only four are endemic, and 26 taxa (35.13%) further reveal broad range endemism. Thus 30 taxa (38.46%) in the family are distributed only within the Himalayan region and the rest account for wides beyond this limit.

Europe holds the maximum number of endemics (45.75%) followedby Soviet Asia (39.93%) which is in tune with their maximun respective strength of the family. In the smaller floristic units eastern Himalaya possesses just 10% endemics .

Rannunculaceae, practically a north temperate element, is distributed in India in the Himalaya. In Kashmir Himalaya, 98 taxa in 17 genera abound with *Ranunculus* as the largest genus (28 taxa) followed by *Thalictrum* (12 taxa). Six genera have just one species each : *Actaea, Caltha , Cimicifuga, Consolida, Pulsatilla, Trollius*. Of the 98 taxa, 20 are endemic (20.40%) and 13.26% reveal specific and 7.41% infra specific endemism. The endemic taxa further display the following local distribution .

Local area	*No.of taxa*
Kashmir	3
Ladakh	3
Kashmir - Ladakh	0
Kashmir - N.Pakistan	8
Ladakh - N. Pakistan	3
Kashmir - Ladakh - N. Pakistan	3
Total	20

Interestingly only three taxa are represented throughout the natural area and the rest lie still localised in their distribution . Among the wides, 20 taxa (25.64%) further extend only upto the Himalayan limit as shown below:

Kasmir Himalaya with	*No.of taxa*
Western Himalaya	12
Western and Central Himalaya	4
Eastern Himalaya	4
Total	20

Thus the family possesses a fairly good proportion of restricted taxa (40.71%) in the study area and the extent of endemism in the neighbouring regions is also quite revealing. Some genera present interesting distribution patterns: aconitums show maximum dominance in Soviet Asia (51 taxa) than Europe (14 taxa) but reverse holds true in as far as the endemic percentage is concerned (33.33 and 85.71, respectively). Significantly 58.28% of the endemic aconitums of Soviet Asia are either recently described new species or new combinations. In Kashmir Himalaya as well, the endemic percentage is no less significant (50%). Genera like *Actaea, Ceratocephalus* and *Cimicifuga* are more or less widely distributed in view of their wider adaptability and are neither represented by endemics in Kashmir Himalaya nor in any of the adjoining areas! But *Consolida* and *Pulsatilla* display 100% endemism in the study area owing to their relatively narrow ecological amplitude.The endemics in *Adonis (25%,) Caltha (14.28%)* and *Paraquilegia*

(40%) are known only in Europe, Soviet Asia and Afghanistan respectively; while *Clematis,* despite being well represented in Soviet Asia, possesses insignificant endemics (5.55%) there. In *Trollius,* the lone species in Kashmir Himalaya *(T. acaulis)* is a broad range endemic. Two weedy genera *Anemone* and *Ranunculus* are both characterised by lower endemism (18.88% and 21.42%, respectively) in the study area. *Aquilegia*(Fig 24), howerver, reveals maximum endemicity (72%)in Europe followed by Kashmir Himalaya (40%) while for *Delphinium* the maxium endemics occur in Soviet Asia (61.72%) and insignificant in Europe (9.67%). *Callianthemum* reveals endemism only in Europe (100%) while *Thalictrum* despite being meagerly represented in Afghanistan (5 taxa) has 40% endemics. A comparative account of the incidence of percentage endemism in some genera of Ranunculaceae in Kashmir Himalaya and India is given below:

Genera	*Kashmir Himalaya*	*India**
Ranunculus	21.42	36.00
Anemone	18.18	43.00
Clematis	0.00	76.00
Thalictrum	8.33	79.00
Delphinium	25.00	71.00
Aconitum	50.00	90.00

* (after Chatterjee, 1939)

The data are quite significant as far as the areas occupied by the two phytogeographical units are concerned. Although Chatterjee's remark that zygomorphy is more favourable towards an active speciation activity and thereby the endemics - is quite justifiable for Kashmir Himalayan genera but is not relevant while considering endemics in the Indian subcontinent. The anomalies are provided by *Clematis* and *Thalictrum* and "it is difficult to see why *Clematis* and *Thalictrum* should show so high a percentage of endemics" (Chatterjee). The infraspecific endemism is also well marked in some genera like *Acquilegia (20%), Anemone (18-18%), Aconitum (10%)* and *Thalictrum* (8.33%) etc thus pointing to active speciation activity.

Berberidaceae incompletely north temperate in distribution with extensions southwards beyond the equator in all tropical sectors except Australia (Good. 1964) is represented by 22 taxa in three genera in Kashmir Himalaya to 35 species in four genera in India. *Berberis* is the only well developed genus (20 taxa) while the other two *Epimedium* and *Mahonia* are monotypic in the study area and restricted in distribution. Eight taxa in *Berberis* (36.36%) are endemic in Kashmir Himalaya and the infraspecific endemism is also quite significant (13.64%). Besides about 50% of the endemics are recently described new species/ varieties-all these facts imply a high speciation activity within the genus in our area. The localised distribution pattern of the endemics is given below:

Local area	*No. of taxa*
Kashmir	4
Ladakh	0
Kashmir - N Pakistan	4
Ladakh - N.Pakistan	0
Kashmir - Ladakh	0
Kashmir - Ladakh - N. Pakistan	0
Total	8

The wides also show a more or less similar distribution pattern:

Kashmir Himalaya with	*No. of taxa*
Western Himalaya	8
Western & Central Himalaya	4
Eastern Himalaya	2
Total	14

Thus Berberidaceae in Kashmir Himalaya is almost entirely restricted in distribution as none of our taxa extend beyond the Himalayan range. Chatterjee recorded 97% endemics in the family in India and while accounting for such a high degree of endemictiy, he opined that gamopeltaous families are more equipped for specific variation as compared to the polypetalous group. A reflection in endemism in the study area and neighbouring regions indicates that *Berberis* harbours maxium

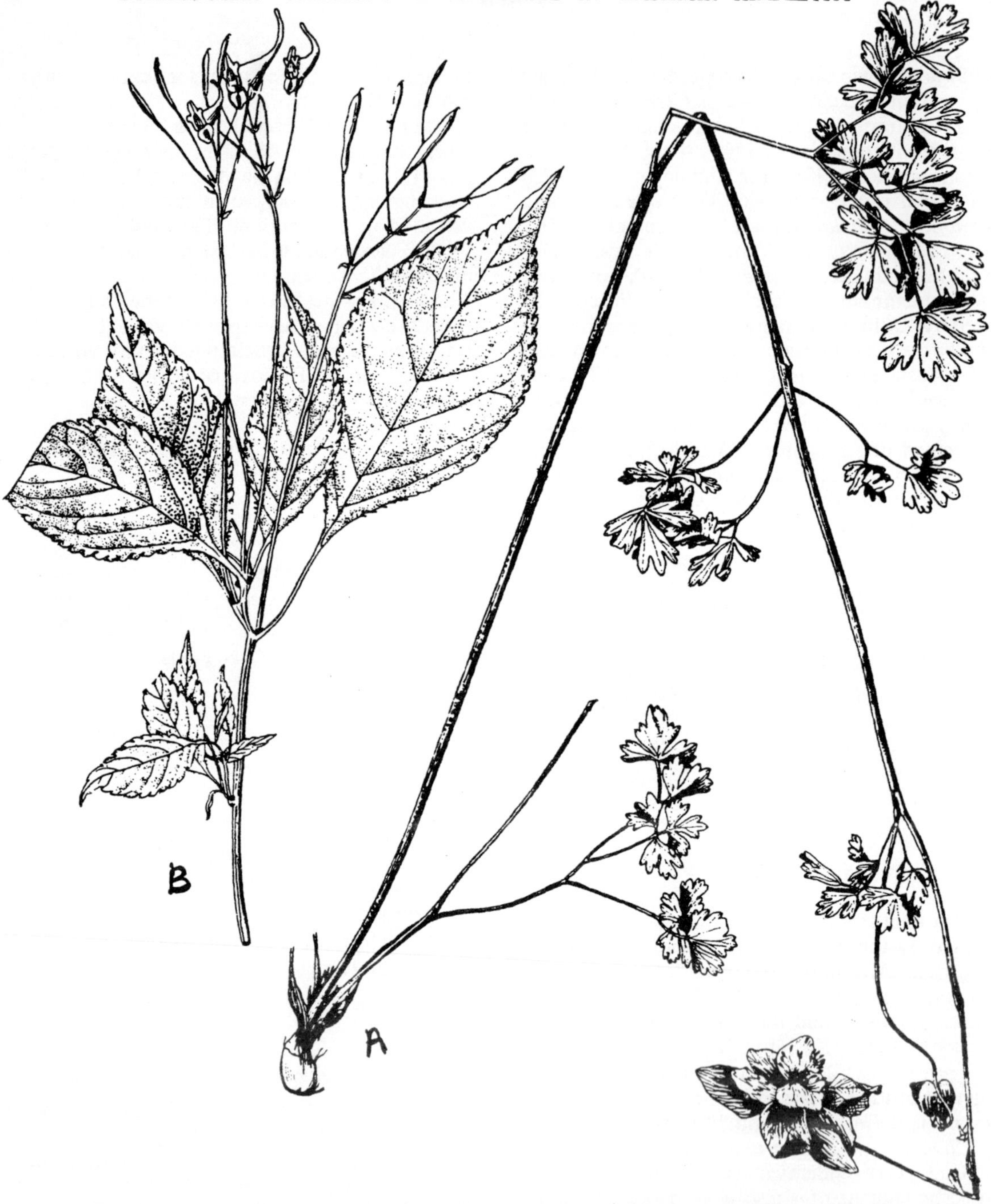

Fig. 24. **A.** *Aquilegia fragrans* Bth. subalpine; of the 5 species of *Aquilegia* in Kashmir 2 are endemic : *A. nivalis* Falc. ex Jackson and *A. pubiflora* Wallich ex Royle.

B. *Impatiens edgeworthii* Hook. f., endemic (Kash) and a very common species of the moist shady places. Of the 14 species of *Impatiens* in Kashmir 6 are endemic.

endemics in the Eastern Himalaya (50%) followed by Kashmir Himalaya (40%) while the major phytogeogrphical units reveal 8 - 25% endmism. Presently , *Epipedium* is known by endemics only in Soviet Asia (20%) while *Mahonia* does not manifest any endemism in these areas. The lone species of *Epimedium* in Kashmir Himalaya *(E. elatum)* is a broad range endemic.

Papaveraceae, a discontinuous taxon between North America to Andes, South Africa and Australia (Good, 1964) is represented by 43 species in seven genera in India with 65.11% endemism (Chatterjee,1939). Nine taxa in four genera occur in Kashmir Himalaya with 11.11% endemism. *Roemeria* and *Hypecoum* are monotypic and widespread. *Papaver* reveals maximum dominance in Soviet Asia (52 taxa) than Europe (31 taxa) but the endemic percentage is almost equal in both the regions. In the smaller phytochoria, endemics in *Papaver* are known only in Pakistan (11.11%) where the lone endemic *P. stewartianum* Jafri and Qaisar is known from the type locality only. *Meconopsis* on the other hand shows strong development in Nepal, Eastern Himalaya and western China (Sino-Japanese region). Two species are restricted in distribution: *M. aculeata* Royle to Western Himalaya and M. *latifolia* Prain to Kashmir Himalaya. Among the adjoining regions, endemics in the genus occur only in the Eastern Himalaya (25%) and Pakistan (50%).

Fumariaceae - allied to Papaveraceae and earlier included in it, is largely confined to north temperate regions. Four genera representing 66 species are known to occur in India (Chatterjee,1939). In Kashmir Himalaya 29 taxa in two genera occur and *Corydalis* is well developed (27 taxa). The percentage of short-range endemics is meagre (13.79%) but the broad endemics together amount to 75.86%, Thus the family remains largely restricted within the Himalayan range. About 75% of the endemics in Kashmir Himalaya are recently described new species/varieties. Chatterjee (1939) also reported strong endemism for the family in India (79%). The localised distribution of broad-range endemics in Kashmir Himalaya is presented below:

Local range: Kashmir Himalaya with	*No. of taxa*
Western Himalaya	4
Western & Central Himalaya	7
Eastern Himalaya	7
Total	18

Corydalis reveals endemism in all the neighbouring areas while *Fumaria* does so only in Europe. Other genera represented by endemics include *Sarcocapnos* (75%) in Europe and *Fumariola* (100%) in Soviet Asia.

A large and cosmopolitan family - **Cruciferae** is chiefly found in temperate regions with cooler climate, In India, it is largely confined to the Western Himalaya and northwest India; a few species also occur in the Eastern Himalaya and Indian Plains but the family is almost lacking in peninsular India in the wild state. In Kashmir Himalaya 134 taxa in 51 genera occur. Jafri. (1973) recorded 31 genera as monotypic in the study area, *Draba* is the largest genus with 20 species followed by *Erysimum* and *Arabidopsis* each with 9 taxa.

Of the 134 taxa, 14 are endemic implying a percentage of 10.41. The endemic taxa further manifest the following localised distribution.

Local area	*No. of taxa*
Kashmir	4
Ladakh	5
Kashmir - Ladakh	1
Kashmir - N. Pakistan	4
Kashmir - Ladakh - N. Pakistan	0
Total	14

Not a single endemic taxon is represented throughout the natural area and ca 57.14% endemics are recently described as new species/ new combinations. Thirtytwo taxa (26.66%) amongst the wides also reveal broad range endemism and their localised distribution is given below:

Local range Kashmir Himalaya with	*No.of taxa*
Western Himalaya	13
Western & Central Himalaya	14
Eastern Himalaya	5
Total	32

In total 46 taxa (34.32%) are confined to the Himalayan range and the rest serve as wides beyond this limit. Chatterjee suggested an endemic figure of 56% in India as being too high for such an invading family. The extent of endemism in Kashmir Himalaya and adjoining phytochoria is quite interesting; *Draba* is more predominant in Soviet Asia (91 taxa) than in Europe (42 taxa) but the endemic percentage is quite opposite (35.16 and 54.76% respectively). In the study area also the genus has maximum number of taxa (20) while the endemic percentage is only 15%. Genera like *Megacarpaea*, *Erophila* and *Microsisymbrium* are represented each by two taxa and 50% endemics in the study area while in the adjoining areas, all these genera are devoid of endemics excepting *Megacarpaea* in Soviet Asia (80%). The endemics in *Arabidopsis* are known only in Kashmir Himalaya and Afghanistan while other three genera revealing endemism in the study area are *Erysimum* (11.11%), *Christolea* (25%) and *Arabis* (22.22%). Soviet Asia contains the maximum number of endemics in all these while Kashmir Himalaya does so in the smaller phytochoria.

A cosmopolitan family - **Crassulaceae** is mostly confined to South Africa, Macronesia and the Mediteranean region. The Macronsian region is largely represented by endemics of *Sempervivum* (Good, 1964). Of the 76.56% endemics in India, 68.75% are encountered only in the Himalayan region (Chatterjee, 1939) which reveals its strong development in this area. Twentyeight taxa in six genera are found in Kashmir Himalaya with 17.85% endemism. Besides nine more taxa (39.13%) discern broad range endemism. In general 14 taxa (50%) are characterised by restricted distribution within the Himalaya.

Sedum (3.87%) and *Sempervivella* (78.59%) are both represented by maximum number of endemics in Europe than Soviet Asia (12.9%,28.57%, respectively). *Sedum* possesses 25% endemics in Kashmir Himalaya, 33.33% in Eastern Himalaya, 14.28% in Afghanistan, 6.66% in Pakistan. *Sempervivella* reveals 66.66% broad - range endemics in Kashmir Himalaya. Some genera in the adjoining areas displaying endemism are: (percentage given):

Pakistan : *Pseudosedum 66.66%.*

Soviet Asia: *Rhodiola 28.57, Pseudosedum 66.66, Rosularia 38.38, Penthorum 50.*

Europe: *Mucizonia 50, Jovibarba 57.14.*

Grossulariaceae, a unigeneric family, is largely distributed in Eurasia, America and Africa. *Ribes* comprises six taxa with 16.66% endemism in Himalaya.

Predominantly a temperate family - **Saxifragacae** is very widespread in America and contains a vast number of widely disjunct genera (Good, 1964). The family comprises 114 species in 17 genera in India with a fairly notable endemic percentage (76.31). In the study area, 33 species in four genera occur. *Saxifraga* and *Chrysosplenium* are exclusively alpine while *Bergenia* ranges from temperate - subalpine - alpine zone. Five taxa (15.75%) are endemic in Kashmir Himalaya while the broad range endemism is quite significant (46.42%). In general 18 taxa (54.54%) are localised within the Himalayan range and the rest extend beyond this. The broad-range endemics are distributed locally as follows:

Local range: Kashmir Himalaya with	*No. of taxa*
Western Himalaya	2
Western and Central Himalaya	7
Eastern Himalaya	4
Total	13

Saxifraga is dominant in Europe (153 taxa) which accords well with its maximum endemic percentage (60.77); followed by Soviet Asia (79 taxa and 44.30% endemics) In the smaller phytochoria both Kashmir Himalaya and the eastern Himalaya head the list in possessing maximum endemics (14.28% each). *Chrysosplenium* reveals greater diversity in Soviet Asia (19 taxa) than Europe (5 taxa) but reverse is true regarding the endemic percentages (40 in Europe and 26.31 respectively). In Kashmir Himalaya, one lone species is endemic (100%) while 33.33% taxa are endemic in the Eastern Himalaya. Among rest of the genera *Bergenia* alone has 50% endemics in Soviet Asia.

Rosaceae, mainly a cosmopolitan family, chiefly inhabits the temperate regions and includes a good number of woody elements. Chatterjee (1939) reported 255 species in 26 genera in India with signifcant endemic percentage (70.19); of this 56.47% occur in the Himalaya.

In Kashmir Himalaya, the endemic figure is 17. 80%, and the local endmics show a high degree of localised distribution; and this together with broad range endemics correspond to 27.29%.

Local area	*No of taxa*
Kashmir	8
Ladakh	2
Kashmir - Ladakh	3
Kashmir - N.Pakistan	5
Kashmir - Ladakh - N. Pakistan	2
Total	20

Alchemilla is dominant both in Soviet Asia (151 taxa) and Europe (120 taxa) with high endemic precentage (75.49 and 78.33 respectively). *Potentilla* is yet another genus possessing greater diversity in both the regions. Same is true of *Craetagus.* Amongst the smaller phytochoria, Kashmir Himalaya represents the centre of major activity for these genera. *Rubus, Spiraea* and *Contoneaster* also reveal maximum endemics in Soviet Asia while in the smaller floristic units, *Rubus* reveals a maximum of 21.74% endemism in the Eastern Himalaya, *Spiraea* (33.33%) in Afghnistan and *Cotoneaster* (16.16%) in Kashmir Himalaya.

Another cosmopolitan family - **Papilionaceae** is chiefly concentrated in northern hemisphere, Chatterjee (1939) recorded 862 species in 12 genera in India with 57% endemism. In the study area 181 taxa in 38 genera occur from temperate to alpine areas. Twenty taxa (11.04%) are endemic in Kashmir Himalaya:

Local area	*No. of taxa*
Kashmir	6
Ladakh	4
Kashmir - Ladakh	1
Kashmir - N.Pakistan	8
Kashmir - Ladakh - N.Pakistan	1
Total	20

Amongst the wides (161 taxa), 22 taxa (13.66%) occur as broad range endemics beyond the Himalayan limit. Apart from short range and broad range endemics, 25 taxa are restricted to the Himalaya and the Punjab (India, Pakistan). The localised distribution of broad range endemics is presented below.

Local range; Kashmir Himalaya with	*No. of taxa*
Western Himalaya	12
Western & Central Himalaya	6
Eastern Himalaya	4
Total	22

Caragana is poorly represented in Europe (3 taxa) compared to Soviet Asia (33 species) and it lacks endemics in Europe and the Eastern Himalaya. *Chesneya* does not reach Europe while of the nine species in Soviet Asia, 77.77% are endemics. In the study area as well, both these genera possess 40% and 50% endemics, respectively. *Oxytropis, Astragalus* and *Hedysarum* are strongly developed in Europe which sychronises well with the endemic percentages. No doubt *Vicia* and *Trigonella* contain almost equal number of taxa in both the major phytochoria but Soviet Asia dominates in endemics. *Lespedeza* like *Chesneya* is not known to occur in Europe and reveals 14.28% endemism in the study area. In the adjoining areas, some genera represented by endemics are the following (percentages given):

Pakistan: *Sophora20, Crotalaria 10, Tephrosia 18.18, Sesbania 50, Colutea 33.33, Lotus 16.16, Onobrychis 11.11, Alysicarpus 12.50.*

Afghanistan: *Cicer* 12.50, *Medicago 11.11, Melilotus 25, Onobrychis 44.44, Onenis 50, Sophora 50.*

Soviet Asia: *Ammothamnus 100, Ammodendron 83.33, Thermopsis 50, Genista 68.42,*

Cytisus 58.83, Medicago 10.44, Melilotus 27.27, Trifolium 7.69, Anthyllis 30.77, Lotus 16.66, Erenosparton 100, Smirnowia 100, Colutea 60.15, Calophaca 80, Glycyrrhiza 50, Meristertrapsis 66.66, Ewersmannia 100, Onobrychis 58.06, Alhagi 40, Cicer 21.42, Lathyrus 24.00, Pisum 20.

Europe: *Cytisus 61.00 Chamaecytisus 74.28, Genista 45.16, Chamaespartium 100, Echinespartum 60, Petteria 100, Ulex 58.33, Stauracanthus 33.33, Lupinus 12.50, Calophaca 100, Cicer 33.33, Lathyrus 19.40, Onenis 77, Medicago 13.33, Trifolium 27.86, Derycnium 12.50, Lotus 25, Anthyllis 50, Coronilla 20, Hippecrepis 33.33, Onobrychis 31.70,* and *Ebenus 50.*

Eastern Himalaya: *Butea 100, Desmodium 22.22.*

In Kashmir Himalaya, endemic percentage is relatively lower in most genera but is sizeable together with broad-range endemics (percentages given) as in:

Astragalus 56.60	*Hedysarum 66.66*
Lespedeza 57.14	*Oxytropis 42.85*
Trigonella 62.50	*Vicia 37.50*

The taxa characterised by broad range endemism only in Kashmir Himalaya are (percentages given).

Indigofera 36.35	*Thermopsis 50*
Atylosia 66.66	*Onobrychis 50*
Rhynchosia 50.00	*Stracheya 100*
Desmodium 37.50	

Geraniaceae is 'heavily massed' in South Africa and South America (Good, 1964). In India, it is chiefly developed in the Himalaya. In Kashmir Himalaya, 24 taxa in two genera occur. Only three taxa (12.50%) are endemic and none displays broad range endemism. About 66.66% endemics include the recently described new varieties. Chatterjee (l.c) reported 50% endemics for the family in the Indian region. *Geranium* the dominant genus contains 15% endemics in the area compared to 28.88% in Europe and 16.36% in Soviet Asia. *Erodium,* a pure temperate element, is known by endemics only in Soviet Asia (27.66%) and Europe (45%).

A southern tropical family - **Rutaceae** also occurs in north temperate region. In India 74 species in 24 genera occur with 32.39% endemism (Chatterjee,1939). In Kashmir Hamalaya, only seven taxa in six genrea are known with 14.28% endemism. *Haplophyllum* is the lone genus represented by endemics in the study area (100%) and Afghanistan (33.33%) while among the major floristic units, genera revealing endemics include *Haplophyllum* (50%) and *Ruta* (20%) in Europe; *Haplophyllum* (59.37%) and *Pictamnus* (40%) in Soviet Asia.At family level, Soviet Asia include large number of endemics (52.50%) followed by Europe and Afghanistan (33.33% each). Endemics are absent in Pakistan and the Eastern Himalaya.

Predominantly a family of Asian and African tropics **Balsaminaceae** also occurs in north temperate region . In India , it shows disjunction between the Himalaya and south India. In Kashmir Himalaya , *Impatiens* is the lone genus with 12 taxa mostly inhabiting moist shady areas(Fig 24). Three taxa (25%) are endemic and six (66.66%) further reveal broad range endemism:

Kashmir Hamalaya and	*No. of taxa*
Western Himalaya	2
Western & Central Himalaya	1
Eastern Himalaya	3
Total	6

Significantly ca 75% balsams are restricted to the Himalaya. Chatterjee recorded a very high percentage (91) for the family in India and that a remarkable development of balsams has taken place in the eastern Himalaya and Burma owing to a significant endemic percentage in this area (61.90%). In Soviet Asia the family includes only 10 species with 20% endemism while none of the taxa is endemic in Europe.

Rhamnaceae is chiefly tropical- subtropical. In Kashmir Himalaya it is represented by 14 taxa in six genera with 7.14% endemism, and the lone species of *Rhamella (R.gilgitica* Mansf) is endemic (7.14%). It is known from the type locality only. The remaining genera, *Berchemia, Helinus. Sagrettia, Zizyphus,* and *Rhamnus* lack endemics here.

In the neighbouring areas, endemics occurs in the following genera (percentages given):

Berchemia: Pakistan 33.33, eastern Himalaya 11.05

Rhamnus: Afghanistan 20, Soviet Asia 21.05, Europe 36.36

Frangula: Europe 33.33

Umbelliferae a cosmopolitan family is widely distributed in boreal temperate and subtropical rogions with greater diversity in the Himalayan region (56.13% endemics: Chatterjee, 1939). In Kashmir Himalaya, 96 taxa occur in 41 genera.and 23 genera are monotypic (Nasir 1972), only *Peucedanum* is characterised by endemics (100%). Further twelve taxa (12.50%) reveal short-range endemism and their local distribution is as follows:

Local area	*No. of taxa*
Kashmir	2
Ladakh	1
Kashmir - Ladakh	1
N. Pakistan	6
Lakakh, N.Pakistan	2
Total	12

Apart from the short-range endemics, 40 taxa (47.61%) further reveal broad-range endemism which in total accounts for 54.16% restricted taxa in the family.

Local range: Kashmir Himalaya and	*No. of taxa*
Western Himalaya	24
Western and Central Himalaya	8
Eastern Himalaya	8
Total	40

Clearly most of the restricted taxa range only upto the western Himalaya suggesting a comparatively recent speciation.

Some other genera represented by endemics are the following (percentages given):

Pakistan: *Aegopodium 50, Trachyspermum 66.66, Eriocycla 33.33, Psam mogeton 25, Cortia 50, Ormopterm 100, Trachydium 50, Stewartiella100, Ferula 18.18.*

Afghanistan: *Angelica 50, Bunium 33.33, Johrenia 100, Reutera 100.*

Europe: *Naufraga100, Sanicula 50, Hacquetia 100, Astrantia 57.14, Eryngium 34.61, Petagonia 100, Anthriscus 28.57, Molopospermum 100, Bunium 44.44, Conopdium 57.14, Huetia 33.33, Dethawia 100, Sesali 50, Oenanthe 40, Portenschlagiella 100, Athamanta 41.66, Grafia 100, Xatardia 100, Trochiscanthes 100, Meum 100, Hladnikia 100, Pleurospermum 50, Cachrys 25, Heptapter 100, Trinia 54.54, Apium 40, Petroselinum 50, Ammi 60, Thorella 100, Carum 57.14, Endressia 100, Ferula 10, Ferulago 55.55, Pastinaca 50, Guillonea 100, Laserpitium 57.90, Thapsia 33.33, Daucus 19.04, Psendorlava 50.*

Estern Himalaya: *Heracleum 75.*

A predominant family of northern hemisphere, **Primulaceae** is mostly Himalayan (159 out of 208 taxa occur here), some elements lie in the south. In Kashmir Himalaya, of the 51 taxa in seven genera, *Primula* is the largest (23 taxa) followed by *Androsace* (18 taxa). Glaux, Anagallis, Cortusa and *Samolus* are monotypic lacking endemics. Only eight taxa (15.68%) reveal short range endemism but the broad range endemism is quite significant (46.51%)

Local and Kashmir Himalayan range	*No. of taxa*
Western Himalaya	9
Western and Central Himalaya	7
Eastern Himalaya	4
Total	20

Generally 27 taxa (54.90%) show restricted distribution within the Himalaya; consequently less than 50% species account for the wides beyond the said limit. Chatterjee had recorded a high endemic percentage for the family in the Indian region(85.09%) and of this 76.44% occur in the

Himalaya and this points to the extent of speciation in the family.

Primula and *Androsace* both show high degree of endemicity in the Eastern Himalaya (80% and 75%, respectively) this signifies the maximum development of the family in this area, In Kashmir Himalaya, 26.08% and 11.11% endemics occur in the two genera, respectively. Endemics in *Lysimachia* are known only in the Eastern Himalaya (14.28%) while *Glaux, Anagallis, Cortusa* and *Samolus* are widespread.

Gentianaceae both tropical and temperate reaches arctic region as well (Good, 1964). Chatterjee reported 188 species in 15 genera from India with 77.65% endemics. 62 species in seven genera occur in the study area and only six taxa (9.67%) are endemic but broad range endemism is quite significant (44.64%) being mostly confined to Western Himalaya. At the family level, ca 50% taxa have restricted distribution . Of the 77.65% endemics in India, Chatterjee recorded 47.34% alone in the Himalayan region. *Gentiana* has 37 taxa and *Swertia* 14 taxa in the study area. Though *Gentiana* reveals greater diversity in Kashmir Himalaya (37 taxa) and Pakistan (24 taxa) yet reverse holds **true** as far as endemic percentage is concerned.

Two genera *Centaurium* and *Halena* have widespread distribution. *Gentianella* has a lone species in the study area, which is endemic but *Lomatogonium* possesses 20% endemics while *Swertia* and *Jaeschkea* are both characterised by broad range endemism. In the Eastern Himalaya *Swertia* reveals 42-85 % endemism.

Essentially tropical **Asclepiadaceae** includes 232 species in 49 genera in India with 73.27% endemism (Chatterjee, 1939). In Kashmir Himalaya 22 taxa in 12 genera occur; *Cynanchum* has a maximum of eight taxa followed by *Periploca* (3 taxa) and *Tylophora* (2 taxa); the remaining nine genera are monotypic here. *Cynanchum auriculatum* Royle ex Wight is the lone species endemic to Kashmir Himalaya (4.54%). The broad-range endemics are 33.33%. *Cynacnchum* (16.16%) is also endemic in Afghanistan; *Sarcoslemma* in Pakistan; and *Hoya(50%)* and *Marsdenia* (50%) in the Eastern Himalaya.

A cosmopolitan/sub-cosmopolitan family - **Boraginaceae**, characteristic of both tropical and the temperate regions, is largely represented in the Himalaya in India. In Kashmir 88 taxa in 26 genera occur (Kazmi, 1974): *Pseudomertensia* with 10 taxa is followed by *Cynoglossum* and *Rochelia*, each with seven taxa and seven genera with single species each: *Anoplocaryum, Asperugo, Gastrocotyle, Lepechiniella, Microula, Trichodesma*, and *Trigonotis*.

Of the 88 taxa, 14 are endemic (15.90%) in which 9.11% reveal specific and the rest infraspcific endemism. These endemic taxa further display the following localised distribution.

Local area	*No. of taxa*
Kashmir	4
Ladakh	1
Kashmir - Ladakh	1
N. Pakistan	6
Ladakh - N. Pakistan	2
Total	14

Only two species flourish throughout the natural area and the rest are more restricted in distribution. In total 39 taxa (44.31%) are characterised by range restriction within the Himalaya. Besides 25 taxa (33.78%) amongst the wides manifest the following broad range endemism:

Local range: Kashmir Himalaya and	*No.of taxa*
Kashmir	13
Western & Central Himalaya	7
Eastern Himalaya	5
Total	25

Five genera represented by endemics in the study area are: *Pseudomertensia (60%), Cynoglossum (28.57%), Mattiastrum (20%), Eritrichium (16.66%),* and *Lindelofia* (20%), The first three genera though gregarious in Pakistan, yet have few endemics there. *Lindelofia* has endemics in Soviet Asia (66.66%). In *Hackelia* endemics occur only in Kashmir Himalaya (50%) while the remaining genera lack endemics here.

Labiatae, a cosmopolitan family centred in the Mediterranean region, has 419 species in 69 genera in India with 62% endemism (Chatterjee, 1939). Of the 127 taxa in the study area, 8.66% show short-range and 32 (27.82%) broad-range endemism. In general, taxa (35.83%) are restricted in distribution within the Himalaya and the rest extend beyond this limit. *Nepeta* is the largest genus (37 taxa) followed by *Scutellaria* (8 taxa) in the study area while 20 other genera are monotypic.

The broad-range endemism is given below:

Local area: Kashmir Himalaya and	*No.of taxa*
Western Himalaya	24
Western and Central Himalaya	6
Eastern Himalaya	2
Total	32

Remarkably 75% of broad-range endemics do not extend beyond the Western Himalaya. Genera like *Salvia, Nepeta, Scutellaria* and *Phlomis* show endemism in Afghnistan and Himalaya excepting Pakistan where only *Scutellaria* and *Nepeta* include endemics and in Kashmir Himalaya, most of the taxa in the two genera are broad range endemis, (57.14% and 50% respectively). Same holds true for *Salvia* (57.14%) and *Phlomis* (100%). Some other genera represented by the broad range endemics include: *Ajuga* (50%), *Elsholtzia*(Fig 25) (57.14%), *Roylea* (100%) and *Stachys* (14.28%) etc. In Soviet Asia most of these genera are richly endemic. The other genera known to possess endemics in the adjoining areas are the following (percentages given):

Afghanistan: *Eriophyton 100, Hymenocrater 33.33, Lagochilus 5, Leonurus 50, Mentha 20, Micromeria 50, Origanum 33.33, Teucrium 25, Thymus 28.87, Ziziphora 25.*

Soviet Asia: *Teucrium 33.33, Marrubium 26.66, Lagopsis 25, Sideritis 70, Lophanthus 25, Dracocephalum 25.71, Kudrjaschevia 25, Hypogomphia 50, Pseuderemostachys 100, Neustruevia 100, Lamium 25, Erianthera 100, Leonurus 16.66, Logochilus 40.74, Otestegia 100, Hallota 33.33, Betonica 57.14, Schraderia 50, Perovskia 83.33, Ziziphora 36.36, Satureia38.46, Micromeria 25, Acinos 25, Hyssopus 55.55, Origanum33.33, Thymus 74.23, Lycopus 24.28, Mentha 25, Elsholtzia 33.33.*

Eastern Himalaya: *Ajuga 66.66, Isodon 12.50, Phlomis 66.66, Pogostemon 33.33, Salvia 33.33, Scutellaria 33.33.*

Chiefly cosmopolitan **Scrophulariaceae** is found in both the tropical and temperate regions. In India the maximum diversity is encountered in the Himalaya. Chatterjee recorded 274 species in 54 genera from India with 50.73% endemics. In Kashmir Himalaya 120 taxa in 26 genera occur. *Pedicularis* is the largest genus (31 taxa) followed by *Veronica* (26) while 15 genera are monotypic and lack strict endemism in the study area: *Antirrhinum, Bramia, Buchnera, Centranthera, Chaenorrhium, Depatorium, Lancea, Leptorhabdes, Limosella, Mimulus, Picrorhiza, Scoparia, Striga, Torenia* and *Wulfenia.* Of the 120 taxa, 39 are endemic (32.50%) and reveal the following localised distribution :

Local area	*No. of taxa*
Kashmir	9
Ladakh	8
Kashmir - Ladakh	6
Kashmir - N.Pakistan	13
Kashmir-Ladakh-N.Pakistan	3
Total	39

Amongst the wides, 23 taxa (28.39%) further exhibit the following broad range distribution:

Local range Kashmir Himalaya	*No. of taxa*
Western Himalaya	13
Western & Central Himalaya	5
Western Himalaya	5
Total	23

The extent of isolation is will marked in Scrophulariaceae. Both the short as well as broad range endemics, and ca 62 taxa (51.66%) are restricted in distribution in the Himalaya. However genera with endemics, in the study area are: *Euphrasia (83.33%), Lagotis (75%), Pedicularis (35.48%), Veronica (30.76%), Kickxia (25%)* and *Scrophularia* (12.50%).

Fig. 25 a) ***Companula cashmeriana*** Royle, together with ***benthamii*** Wall is restricted in distribution and do not extend beyond Afghanistan; 3 species are endemic : ***argyrotricha*** Wall ex DC and ***aristata*** Wall ex Royle (W & Himal), *colorata* var *tibetica* Hook f & T (Kash-Pak).

b) *Elshotzia densa* Bth, temperate-alpine, 6 species occur in Kashmir, *eriostachya* var *pusilla*

In Afghanistan *Euphrasia* reveals 100% endemics and *Pedicularis* and *Veronica* 40% each. The latter are poor in endemics in Pakistan and Afghanistan. *Scrophularia* has just 12.50% endemics in Kashmir Himalaya while 21.42% in Pakistan; 37.50% in Afghanistan and 100% in the Eastern Himalaya. Considering the broad range endemics the genus in the study area is exclusively confined within the Himalayan range. Besides afore-mentioned genera, broad range endemics are also known to occur in *Picrorhiza* (100%), *Verbascum (25%)*, *Veronica (11.53%)*, *Pedicularis (29.03%)* and *Linaria* (33.33%) in Kashmir Himalaya. Of other genera showing endemism are *Linaria* (18.18%) in Afghanistan and *Oresolon* and *Picrorhiza* (100% each) in the Eastern Himalaya.

Orobanchaceae is a disjunct family of the north temperate zone, South America and Australia. It has 29 species in seven genera with 41.37% endemism in India.Only three taxa in two genera occur in the study area with 30.76% endemism. *Orobanche* has 12 taxa in Kashmir Hamalaya with 33.33% endemics while *Xylanche* is monotypic and widespread in distribution. In neighbouring Afghanistan only *Cistanche* reveals 50% endemics.

A large and subcosmopolitan family - **Rubiaceae** is essentially tropical in distribution and very well represented in continental India, Assam, Burma and the sub- tropical regions of the Himalaya. Chatterjee regarded its main centre of development in South India (Ceylon) and southern Burma. Further he recorded 555 species in 68 genera from this region in which 368 species are endemic.

In Kashmir Himalaya, 35 taxa occur in 14 genera: *Galium* is largest genus (13 taxa) followed by *Rubia* (5 taxa), and eight genera are monotypic.

Galium alone is represented by endemics in Kashmir Himalaya (7.69%), while *Rubia* does so in Pakistan (12.50%). The former reveals maxium endemics in Europe (62.33%) which accords well with its greatest diversity (154 taxa) in that region. In Afghanistan and the eastern Himalaya, this genus is rather poorly represented by endemics (7.69% and 14.28% respectively). Other genera including endemics in some of the neighbouring areas are:

Afghanistan: *Rubia 14.28%, Asperula 10%, Gaillenia 28.57%,*

Europe: *Crucianella 12.50%, Asperula 67.94%, Cruciata 28.57%, Valantia 33.33%.*

Eastern Himalaya: *Lasianthus, Ixora, Leculia, Psychortia (100% each), Ophiorrhiza 80%, Rubia, wendlandia (150% each) Neanotis 25%.*

A widespread family - **Caprifoliaceae** is chiefly temperate and in India it shows strong development in the Himalaya, Chatterjee recorded 56 species in eight genera in India with a notable endemic percentage (76.78%). *Lonicera* is the only well developed genus with (22 taxa) 22.72% endemics in Kashmir Himalaya; five (17.85%) reveal short-range and nine (39.13%) broad-range endemism, and this in total corresponds to 50% restricted taxa in the family. Endemics in Afghanistan are 6.25%, Eastern Himalaya 16.66% and Europe 26.08%. *Viburnum* neither includes endemics in Kashmir Himalaya nor in any adjacent area. However, 66.66% taxa are broad range endemics in Kashmir Himalaya.

Valerianaceae is one of the discontinuous elements being distributed in North & South America and parts of Malaysia. It is better represented in northern hemisphere. In India 20 species in four genera occur with a remarkable endemic percentage (80) and in the Himalayan region alone 50% taxa are endemic (Chatterjee l.c).

In Kashmir Himalaya 14 taxa in two genera flourish, three (21.42%) being endemic. The broad-range endemism amounts to 27.27%. In total six taxa (42.85%) in the family are restricted in distribution within the Himalayan belt.

Valeriana the only genus with a significant percentage of endemism in Europe (48.14) has ca 27.27, endemism in Kashmir while in *Valerianella* and *Centranthus* endemics are knowm only in Europe (9.09 and 43.75%, respectively).

A monotypic family, **Morinaceae,** is mainly distributed in Southeast Asia and Europe. *Morina* reveals 50% endemism in Kashmir Himalaya.

A cosmopolitan/subcosmopolitan family **Campanulaceae** mostly confined to temperate regions is represented by 71 species in 10 genera in India; and 14 taxa in three genera in the study area. The endemic percentage being 66.19% in the former and 14.28% in the latter. *Campanula* comprises nine taxa with 22.22% endemics in the study area(Fig 25). Among the wides eight taxa (66.66%) show broad- range endemism. In general, the Kashmir Himalaya taxa are largely restricted within the Himalaya (71.42%) and only less than 30% extend beyond this limit as wides. Except for the eastern Himalaya, *Campanula* reveals endemics in all the adjoining floristic units. It dominates both in number and endemic percentage in Europe (204 and 66.12, respectively) compared to 13.33% endemics in Pakistan and 15.38% in Afghanistan. But *Codonopsis* is known by endemics only in the Eastern Himalaya (66.66%) and does not reach Europe. In the study area, two of the three species reveal broad range endemism. *Asyneuma* possesses endemics in Europe (50%) and Afghanistan (33.33%) while both the representaives of the genus in Kashmir Himalaya are restricted in distribution. Other genera characterised by endemics in Europe include: *Azorina (100%), Adenophora (100%), Symphyandra (50%), Trachelium (42.85%), Petromorula (100%), Phyteuma (80%), Phyxopleris (100%) Wablenbergia (50%), Edraianthus (72.72%), Jasione (50%).*

A large and ubiquitous family **Compositae** constitutes one of the dominant groups in flora and ranges widely from tropical regions to the high alpines. Chatterjee (1939) recorded 6967 species in 126 genera from India with 52% endemism; the specific content being almost equal in south India and the Himalaya. In Kashmir Himalaya of the 392 taxa in 78 genera *Taraxacum* (70 taxa including V. Soest's species), *Saussurea* (35 taxa) and *Artemisia* (32 taxa) are well represented followed by *Lactuca* (17 taxa), *Senecio* (14 taxa)(Fig 26), and *Erigeron* and *Inula* (12 taxa each); besides 31 genera are monotypic mostly wide in distribution.

Of the 391 taxa, 108 are endemic (27.61%) and have localised distribution:

Local area	*No. of taxa*
Kashmir	51
Ladakh	13
Kashmir - Ladakh	14
N. Pakistan	22
Ladakh - N. Pakistan	1
Kashmir - Ladakh - N. Pakistan	7
Total	108

Ca 50% taxa are confined to Kashmir while among the rest, only seven taxa are common throughout the natural area. Apart from short-range endemics, 74 taxa (26.14%) further reveal varying degree of range restriction within the Himalayan limit as under.

Local range: Kashmir Himalaya and	*No. of taxa*
Western Himalaya	29
Western and Central Himalaya	34
Eastern Himalaya	11
Total	74

Thus 182 taxa (46.54%) in the study area are characterised by range restriction within the Himalaya, a figure which is of great relevance especially while considering the highly efficient dispersal capacity of the family. Endemism in Kashmir Himalaya and the adjoining phytochoria show peculiarities.

Codonocephalum, Dubyaea and *Soroseris* are monotypic in Kashmir Himalaya and support 100% endemics. *Anaphalis, Artemisia, Chrysanthemum, Erigeron, Inula, Juninea, Lactuca, Saussurea, Scorzonera. Senecio* and *Tanacetum* are largely represented by endemics in almost all the phytogeographical areas. Some genera, known by endemics in the study area include: *Chondrilla (33.33%) Crisium (14.28%), Cremanthodium (25.00%), Gnaphalium (20.00%) Hieracium (16.16%), Launaea (20.00%), Leonotopodinum (28.57%) , Taraxacum, 77.14%), Waldheimia (20.00%),* and *Youngia (33.33%). Artemisia* and *Senecio* though dominant in Europe (71 and 104 taxa respectively) yet reveal almost equal incidence of endemism in all the adjoining areas. *Charysanthemum,* however, presents a reverse case - minimum diversity in Europe (2 taxa) and maximum endemic percentage in Afghanistan (46.66%). *Saussurea* exhibits minimum development in Kashmir HImalaya (35 taxa) with 31.42% endemics. However, while considering short- and broad-range endemics together the figure reads 74.28%. Other genera with broad range endemics in the study area include *Carduus, Gerbera, Olgaea, Phagnalon (100% each), Tricholapis (80%), Scorzonera, Tanacetum, Doronicum,* ***Chrysnthemum (66.66%), Ligularia, Tragopogon, Ainsliaea (50% each), Cousinia (40%), Anaphalis (37.50%) Senecio (36.36%), Inula (33.33%), Brachyactis, Psychorogeton (33.33% each) Aster (30%), Lactuca (28.57%), Heteropapus (25%),*** and *Crepis (12.50%).* The remaining 39 genera in the

Fig. 26 : *Senecio jaquementianus* (Decne) Bth ex Hook f, an alpine growing in shady slopes in Kashmir Himalaya of the 13 species in Kashmir, 8 are alpine-subalpine, 5 endemics : *levingii* Cl (Kash), *chenepodifolius* DC, *chrysanthemoides* var *sisymbriiformis* (DC) Hook f (Kash-Pak), *tibeticus* Hook f (Kash-Pak).

study area are fairly widespread, In the adjoining areas some other genera represented by endemics are (percentages given):

Pakistan: *Centaurea 16.66, Cousinia 11.53, Launaea 20.00, Taraxcum 31.50, Wendelboa 100.*

Afghanistan: *Anthemis 20, Aster 20, Atractylis 100, Blumea 66.66, Carduus 33.33, Centaurea 27.27, Chartolepis 100, Chondrilla 58, Cirsium 23.07, Echinops 50, Gnaphalum 33.33, Hieracium 33.33, Ixeris 100, Ligularia 50, Olicochaela 50, Perlya 50, Picris 50, Plucaria 38.46, Rhapontium 50, Tragopogon 37.50, Varthemia 33.33.*

Europe: *Bellis 33.33, Bellium 66.66, Aster 20, Filago 9.09, Logfia 16.66, Evax 40, Omalotheca 50, Helichrysum 29.16, Antennaria 50, Phagnalon 50, Dittrichia 25, Pulicaria 20, Jasonia 50, Buphathalmum 100, Telekia 50, Santolina 27.27, Anthemis 52.32, Achillea 46.26, Matricaria 18.18, Chamimilla 25, Phalacrocarpum 100, Leucanthemopsis 38.46, Prolongoa 100, Lepidophorum 100, Hymencstemma 100, Leucanthemum 42.10, Plagius 100, Mananther 100, Petasites 23.07, Homogyne 100, Adenostylis 60, Arnica 16.66, Doronicum 57.14., Cartina 32, Atractylis 50, Echinops 18.69, Berardia 100, Staehelina 60, Carduus 32.85, Cirsium 56.33, Ptilostemon 45.45, Lamyropsis 50, Onopordum 38.46, Cynara 42.85, Palaeocyanus 100, Cheirolophus 33.33, Serratula 36.36, Leuzea 44.44, Cyanopsis 100, Phalacrachena 100, Centaurea 50.9, Wagenitizia 100, Carthamus 18.18, Carduncellus 66.66, Rothmaleria 100, Hymenonema 100, Tolpis 40, Hyoseris 20, Hypochoeris 33.33, Leontoson 48.07% Picris 41.17, Tragopogon 36.84, Reichardia 25, Launaea 33.33, Aetheorhiza 33.33, Sonchus 7.14, Cicerbita 50.00, Taraxacum 32.56, Chondrilla 50, Calycocorsus 100, Crepis 48, Hispidella 100, Andryala 40, Hieracium 59.10.*

Eastern Himalaya: *Adenoslemna 50, Carpesium 25, Cicerbita 50, Conyza 33.33, Cremanthodium 100, Gerbera 50, Mikania 50, Vernonia 16.66.*

The overall endemic percentage in Kashmir Himalaya and some neighbouring areas is as follows:

Kashmir Himalaya	27.61
Afghanistan	38.51
Pakistan	7.49
Europe	60.28
E. Himalaya	14.49

MONOCOTYLEDONS

Liliaceae with maxium diversity in warm temperate and tropical regions includes 33 species in 16 genera in the study area. Two species (6.06%) are endemic and five (16.12%) have restricted distribution. Seven species (21.21) are confined to the Himalaya and the rest extend as wides beyond this limit. *Gagea* is the lone genus with 28.57% endemics in Kashmir Himalaya compared to 48.64% in Soviet Asia. Genera with broad range endemics in the study area are: *Lilium (100%), Notholirion (100%), Polygonatum (33.33%)* and *Tulipa (33.33%)*. Among the adjoining areas, genera displaying endemism are (in percent):

Afghanistan: *Eremurus 5.88, Polygonatum* 25.

Pakistan: *Asparagus 9.09, Fritillaria 33.33.*

Soviet Asia: *Narthecium 100, Tofieldia 25, Stenanthium 100, Veratrum 28.57, Merendera 37.50, Asphodilus 60, Eremurus 65.21, Lillum 20, Korolkowia 100, Rhinopetalum 80, Fritillaria 57.69, Tulipa 71.42, Erythronium 50, Scilla 29.4, Ornithogalum 45.83, Bellevalia 82.35, Hyacinthus 33.33, Hyacintella 50, Muscaria 38.88, Asparagus 25, Polygonatum 29.41.*

Eastern Himalaya: *Aletris 33.33, Chlorophytum 33.33, Nothlirion 100, Peliosanthes 50, Polygonatum 25, Similacina 40, Smilax 33.33,* and *Tofieldia* 100.

The overall endemic percentages are:

Kashmir Himalaya	6.06
Eastern Himalaya	21.27
Pakistan	3.27
Afghanistan	2.77
Soviet Asia	46.93

A temperate genus extending into Africa in the south, *Allium* comprises 21 taxa in Kashmir Himalaya with 4.76% endemism. The broad range endemism amounts to 25%. Six species (28.57%) reveal varying degrees of range restriction within the Himalaya and the rest extend beyond this limit. The genus has 5.88% endemics in Pakistan; 14.28% in Afghanistan, 48.68% in Soviet Asia; but none in the Eastern Himalaya.

Orchidaceae, a cosmopolitan family better represented in the tropics than temperate regions, includes 39 species in 22 genera in Kashmir Himalaya with 7.69% endemics. Among the wides 11 taxa (30,55%)further represent range endemism. Thus 14 taxa (35.89%) reveal varying degrees of range restriction within the Himalaya and the rest serve as wides beyond this limit. The extent of endemism reveals the following interesting results.

Epipactis possesses 33.33% endemics in the study area and none in the adjoining areas. *Neottia* is totally restricted in its distribution, with 66.66% endemics in Kashmir Himalaya and the rest accounts for broad range (west Himalaya) endemism. Some other genera represented by broad range endemics in the study area include: *Habenaria (25%) Gymnadenia (100%), Gastradia (100%), Epipogium (50%), Calanthe (50%), Cypripedium (100%), Dactylorhiza (100%), Herminium(25%),* and *Malaxis (50%).* In the neighbouring areas, genera including endemics are: (percentage given):

Pakistan: *Eulophia 33.33.*

Eastern Himalaya: *Acriopsis 100, Bulbophyllum 35.71, Calanthe 100, Geologyne 16.66, Cremastra 100, Cymbidium 16.66, Dendrobium 10, Eria 14.28, Gastrochilus 33.33, Goodyera 20, Malaxis 40, Otochilus 50, Pleione 33.33, Thrixspermum 100, Tripularia 100.*

South Asia:*Listera 25, Habenaria 50, Platanthera 12.50, Orchis 11.11, Himantoglossum 100, Ophrys 33.33.*

The total endemic percentages for the family are:

Kashmir Himalaya	7.69
Eastern Himalaya	9.16
Pakistan	2.67
Soviet Asia	9.02
Afghanistan	0.00

In Kashmir Himalaya **Cyperaceae** a cosmopolitan family is represented by 156 species in nine genera. *Carex* and *Cyperus* are the only two notable genera followed by *Kobresia*; while *Eriophorum* , *Cladium* and *Bubostylis* are either mono or bitypic and *Carex (16.20%), Eleocharis (18.57%),* and *Kobresia (100%)* are represented by endemics. Besides 15 taxa (10.71%) amongst the wides manifest broad range endemism.31 taxa (19.87%) have restricted distribution within the Himalaya and the remaining taxa extend beyond this limit. In the adjoining areas endemics are represented by:

Pakistan: *Carex 3.84, Cyperus 3.34.*

Soviet Asia: *Carex 10.20, Kobresia 12.50, Fimbristylis 33.33, Heleocharis 3.03, Scirpus 15.79, Kyllinga 30, Cyperus 7.14, Pycreus 30.*

Afghanistan: *Carex 3.55%, Eleocharis 33.33%, Scirpus 14.28.*

Eastern Himalaya: *Carex '16.66, Kobresia* 9.09.

The total endemic percentages for the family are: Kashmir Himalaya 10.32, E. Himalaya 6.52, Pakistan 2.61, Soviet Asia 10.09, Afghanistan 6.75.

A large and cosmopolitan family **Gramineae** comprises 314 species in 108 genera in Kashmir Himalaya with 4.77% endemism. In addition to this 39 taxa reveal broad range endemism whose localised distribution is presented below:

Local range, Kashmir Himalaya and	*No. of taxa*
Western Himalaya	20
Western and Central Himalaya	9
Eastern Himalaya	10
Total	39

In general 54 taxa (17.19%) in the family exhibit range restriction within the study area, a figure which does not seem insignificant while considering the weedy nature of the family.

Rosteria (22.22%), Schizachyrium (100%) and *Digitaria (12.50%)* have endemics only in the study area: *Puccinellia* in Kashmir Himalaya (28.57%) and Pakistan (25.00%); *Calamagrostis* in Kashmir Himalaya 17.64%), Pakistan (13.13%) and Afghanistan (20%); *Poa* in Kashmir Himalaya (3.44%), Eastern Himalaya (15.38%), and Soviet Asia (61.6%); and *Festuca* in Kashmir Himalaya (20%), Pakistan (6.25%), Afghanistan (28.57) and Soviet Asia (43.13). Besides some other genera represented by endemics in the adjoining areas include *Duthiea 50, Stipa 11.11* and *Elyhordeum 100.* The following genera reveal broad-range endemism in the adjoining areas (percentages given):

Pakistan: *Danthonia 50, Duthiea 100, Orinus 100, Muehlenberghia 33.33, Isachne 100, Pennisetum 33.33, Cymbopogon 16.66, Saccharum 25, Themeda 100, Festuca 25, Catabrosella 100, Hyalepoa 100, Poa 27,58, Puccinellia 28.57, Melica 20, Agrostis 27.27, Calamagrostis 11.11, Hierochloe 100, Phleum 33.33, Stipa 13.63, Trikeraria 100, Hordeum 14.28, Elymus 5.88.*

Afghanistan: *Bromus 9.09, Oryzopsis 28.57, Roegneria 33.33, Stipa 12.50.*

Soviet Asia: *Apluda 100, Andropogon 50, Eriochlon 50, Echinochloa 28.56, Aristida 40, Ptilagrostis 25, Stipa* 49.12, Oryzopsis 50, Timouria 100, Crypsis 40, Alopecurus 34.48, Limnas 100, Arctagrostis 50, Agrostis 44.82, Deschampsis 38.46, Trisetm 27.27, Avena 5.88, Ayenastrum 42.85, Phragmites 33.33, Diplachne 20, Eragrostis 22.22, Keeleria 58.62, Melica 10.52, Briza 16.66, Dactylis 50, Schismus33.33, Gellardiechloa 100, Eremopoa 50,Colpedium 50, Catabrosa 50, Glyceria 31.25, Atropsis 65, Leucopoa 20, Scleropoa 66.66, Schizachne 100, Littledalea 100, Bromus 34.09, Brachypodium 25, Roegneria 52.50, Agrophyrum 56.86, Secale 20, Asperella 33.33, Aneurolenidium 60, Pasathyrostachys 66.66.

Eastern Himalaya: *Agrostis 28.57, Arthraxon 50, Bromus 50, Erianthus 40, Isachne 25, Themeda 33.33, Arundinaria 100, Dendrocalamus 50, Thamnocalamus 100.*

The total incidence of endemism (%) in the family is:

Kashmir Himalaya	4.77
Eastern Himalay	8.76
Pakistan	1.95
Soviet Asia	37.23
Afghanistan	3.79.

GENERA WITH ENDEMICS

Betulaceae: *Betula.*
Boraginaceae: *Arnebia, Cynoglossum, Eritrichum.*
Callitrichaceae: *Callitriche.*
Campanulaceae: *Campanula.*
Chenopodiaceae: *Haloxylon.*
Compositae: *Aster, Artemisia, Anaphalis, Cirsium; Jurinea, Saussurea.*
Cruciferae: *Arabidopsis, Arabis, Braya, Draba, Megacarpaea.*
Cyperaceae: *Carex.*
Dipsacaceae: *Dipsacus.*
Fumariaceae: *Corydalis.*
Gentianaceae: *Gentiana, Gentianella, Lomantogonium, Swertia.*
Geraniaceae: *Geranium.*
Gramineae: *Bromus, Phleum, Melika.*
Labiatae: *Nepeta, Scutellaria.*
Leguminosae: *Astragalus, Caragana, Hedysarum, Oxytrophis.*
Liliaceae: *Polygonatum.*
Orchidaceae: *Herminium.*
Primulaceae: *Androsace, Primula.*
Ranunculaceae: *Aconitum, Anemone, Pulsatilla, Thalictrum.*
Rosaceae: *Cotoneaster, Potentilla.*
Rubiaceae: *Asperula, Galium.*
Saxifragaceae: *Saxifraga.*
Scrophulariaceae: *Euphrasia, Lagotis, Veronica.*
Umbelliferae: *Bupleurum, Heracleum, Aulacospermum.*
Valerianaceae: *Valeriana.*

Abbreviations used

Him, Himalaya (W, western, C. central, E. eastern). Kash , Kashmir, *Karak,* Karakoram. *Kg,* Kirgiz. *Pak,* Pakistan. *Pam,* Pamir, (incld Pamir-Alai) *Khz,* Kazakhistan. *Turk,* Turkeminia. *Uzb,* Uzbekistan, *Jap,* Japan. *Amer,* America, *Eur,* Europe.

Common, not common: this is with respect to the areas discussed in the text!

Betula

B. utilis D Don is the only species in Kashmir and *B. utilis* var *kitamara* alone occurs in Karakoram. There are reports of *B. utilis* occurring in Afghanistan and central China also. The genus is rich in Central Asia and the USSR.

Common
Uzb-Kg . . *procura* Litv, *turkestanica* Litv, *verrucosa* Ehrb.
Kg-Pam. . *korschinskyi* Litv, *alajica* Litv.

Not common
Kg . . . *pubescens* Ehrb, *saposchnikovii* Sukacz, *tianschanika* Rupr.
Pam. . *davurica* Pall, *lanata* (Rgl) V Vassil, *jarmolenkoana* Golosk, *microphylla* (Bge) V Vassil, *nana* L., *ovczinnikovii, schugnanica* (B Fedt)Litv.

Arnebia

Of the four species, *A. benthamii* (Wall ex G Don)IM Johns is endemic (Kash-Pak), *A. guttata* Bge reaches Mongolia.

Common
Turk-Uzb-Kg . . *decumbens* (Vent) Coss et Kral.
-Uzb . . *minima* Wattst *(=linearifolia* DC), *transcaspica* MP.
Uzb- Kg . . *baldschuanica* (Lipsky) Schinshk.
Kg- Pam . . *obovata* Bge, *tibetana* Kurz.
-Kash . . *guttata* Bge.

Not commom
Turk . . *grandiflora* MP
Kash . . *euchroma* (Royal)IM Johns, *griffithii* Boiss.

Cynoglossum (Fig. 27)

Seven species occur here, four are endemic: *microglochin* Bth var *marginatum* Braud (Kash), *thomsonii* (Cl)IM Johns (Kash), *microglochin* Bth (W Him), *petiolatum* (HK f)DC (W,C, Him).

Common

Turk-Uzb-Kg .. *creticum* Mill (=*pictum* Ait).

Uzb-Kg .. *capusii* (Franch) Pazij, *viridiflorum* Pall ex Lehm

Not common

Uzb .. *seravaschanicum* (B Fedt)MP

Kg .. *officinale* L.

Kash .. *glochidiatum* Wall ex Bth, *lanceolatum* Forssk, *zeylanicum* (Vahl)Thunb ex Lehm.

Eritrichum

Five species occur in Kashmir. *spathulatum* (Bth in Royle) CB Cl var *thomsonii* (CB Cl) Kazim is endemic (Ladakh). The genus is not recorded from Uzb and Kg.

not common

Pam .. *dubium* O Fedt, *pamiricum* B Fedt, *pseudostrictum* MP, *subjacquemontii* MP (=*turkestanicum* auct).

Kash .. *canum* (Bth in Royle)Kitam, *fruticosum* Klotz, *nanum* (Vill) Schrad ssp *villosum* (Ldb)Braud, *patens* Decne, (*nanum* is perhaps endemic to Gilgit - Zanskar).

Callitriche

Two species occur in our area: *stagnelis* Scop is endemic and *palustris* L. (=*verna* L.) is common to Uzb-Kg-Khz. C. *hermophrodite* L. is common to Kg and Pam.

Campanula (Fig 25)

Of the eight species reported in Kashmir, *argyrotricha* Wall ex DC, *aristata* Wall ex Roxb (W,C, Him) and *colorata* var *tibetica* HK f et T (Kash-Pak) are endemic.

Common

Uzb- Kg .. *albertii* Trautv, *evolvulacea* Royle, *fedschekoana* Trautv, *glomerata* L, *lehmanniana* Bge.

Not common

Uzb .. *fastigiata* Duf

Kg .. *eugeniae*, *medium* L.

Turk .. *incanascens* Bss (only species !)

Karak .. *colorata* Wall var *tibetica* HK f et T (only species !)

Kash .. *alisnoides* HK et T, *benthamii* Wall, *cashemiriana* Royle, *colorata* Wall. *latifolia* L. *tenuissima* Dunn.

Haloxylon

H. thomsonii Bge et Bss is the sole species and endemic to Ladakh (Astor-Ladakh). There are no records from Kg. Three species *ammodendron* (CAM)Bge, *aphyllum* (Minkv)Iljin and *persicum* Bge occur in Uzb-Pam Alai.

Anaphalis (Fig.28)

Nine species occur in our area: *A, royleana* DC (Kash- Pak), and *royleana* var *concolor* HK f (Kash) are endemic, *nepalensis* (Spreng) Hand-Maz reaches China and *timmua* D Don North America.

Common

Uzb-Kg .. *racemifolia* Franch, *rosea-alba* Krasch, *velutina* Krasch.

Pam-Karak-Kash .. *virgata* Thoms ex Cl.

-Kash ..... *contorta* (D Don)Hk f.

Not common

Uzb .. *seravaschanica* (Winkl)B Fedt

Kg .. *deoauperata* Boriss

Karak. . *nubigena* DC

Kash .. *adnata* DC, *busua* (Buch-Ham)Hand-Maz, *nepalensis* (Spreng) Hand-Maz, *triplinervis* (Sims)Cl

Artemisia (Fig 28)

Of the 31 species found is Kashmir, five are endemic:

Kash-Pak .. *amygdalina* Decne, *roxburghiana* var *grata* (Wall) HK f.

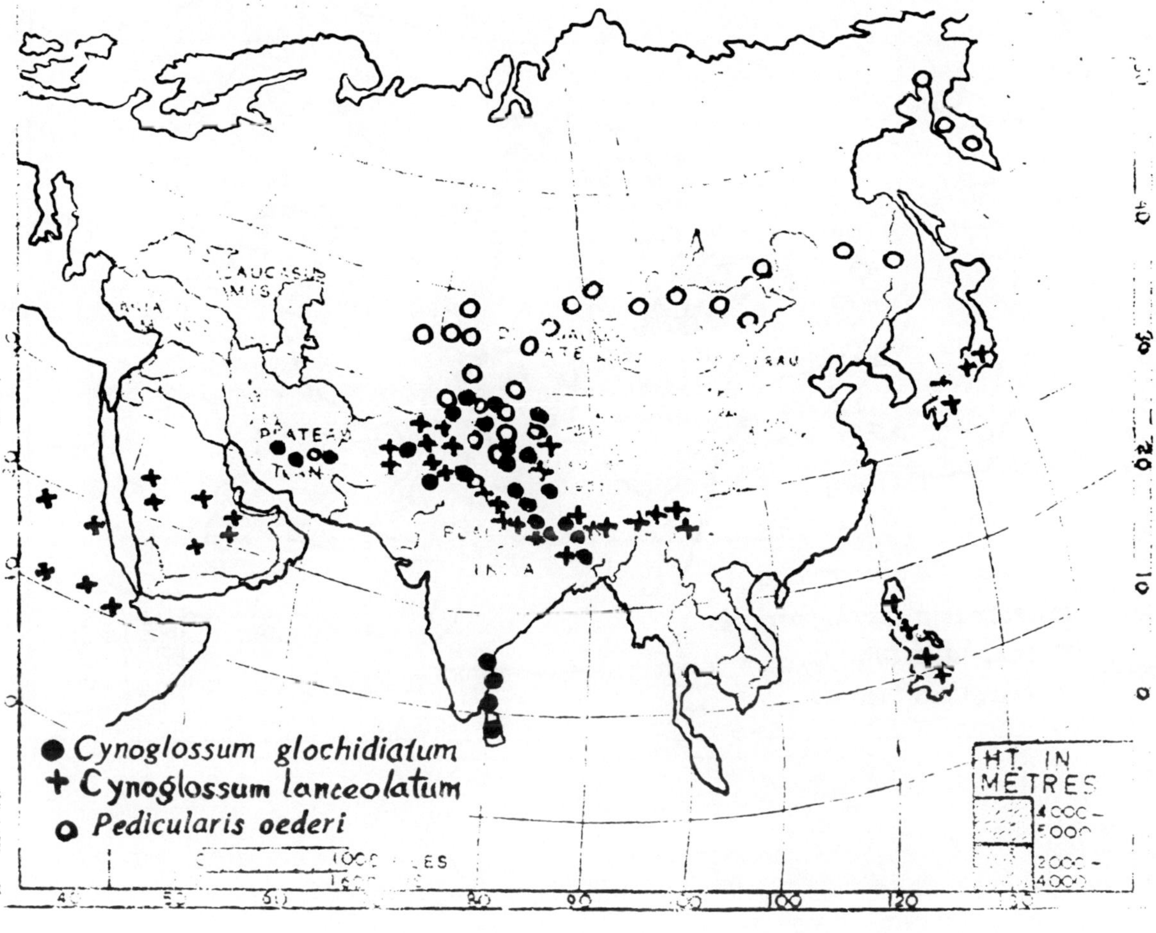
Cynoglossum glochidiatum
Cynoglossum lanceolatum
Pedicularis oederi
INDIA
HT. IN METRES

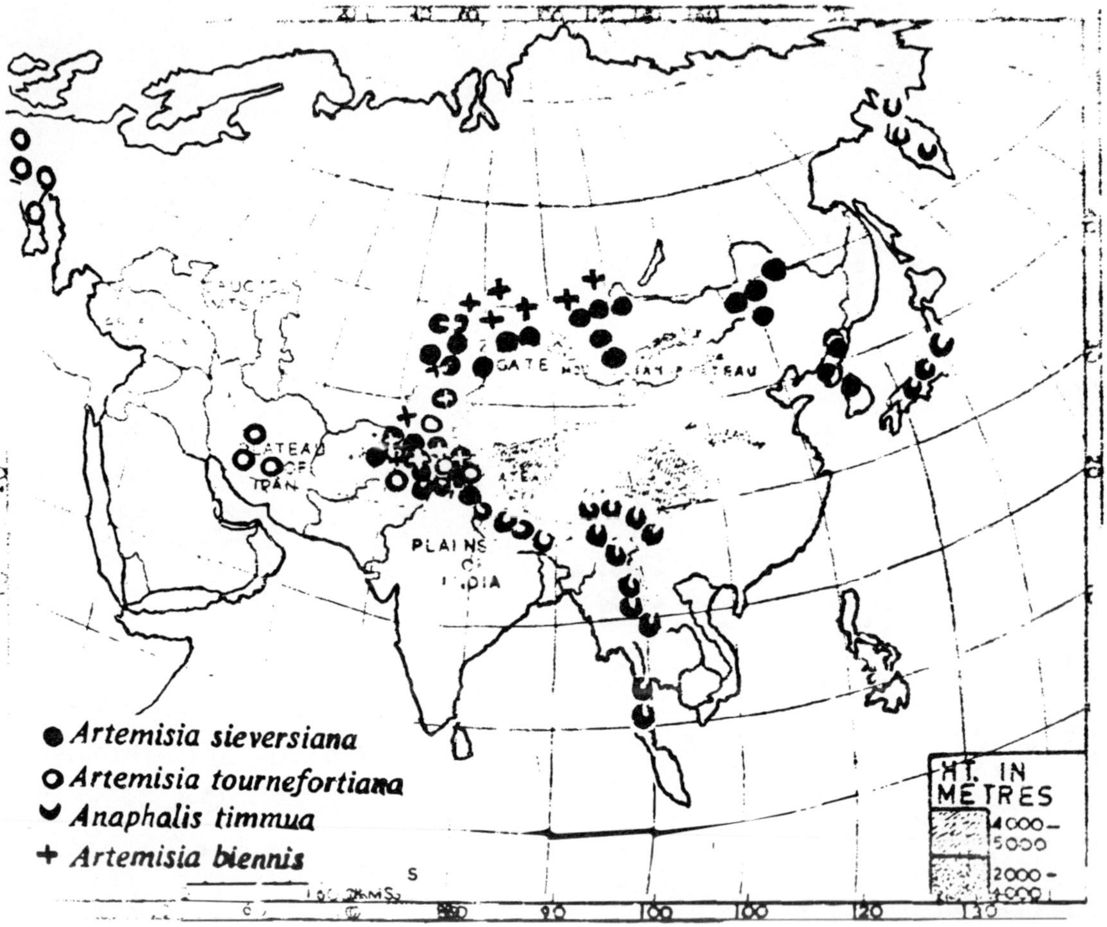

Fig. 28 : *Anaphalis* abounds in Asia Europe and America. None of our elements extend westwards beyond Iran. *Artemisia* essentially a temperate genus extends into tropical mountains but even our widespread species do not extend into SE Asia. A number of species common to Japan and China do not occur in Central Asia.

Ladakh-Pak . . *campbellii* Spreng
Ladakh . . *salsoides* var *walbyi* (Hemsl et Pearson)Osten
W Him . . *salsoides* var *paniculata* Hk f.

THE ALTITUDINAL DIVERSITY

Kashmir - Pakistan (cf RRS).

500m . . . . . *Salsoides* Willd† var *wellbyi* (Hemsl et Peterson)Osten
1000-3000m . . . *roxburghiana* Wall ex Besser.
-3600 . . . *parviflora* Roxb.
1600-3000 . . . *absinthium*+m L, *roxburghiana.*
-3300 . . . *roxburghiana* var *purpurescens* (Jacq ex Besser)Hk f.
-4000 . . . *laciniata* Willdm
-4700 . . . *persica* Bss.
1700-3000 m . . *vestita* Wall ex DC, *? incisa* Pamp*.
2000-4500 . . . *sieversiana* Ehrb ex Willdm+
-3800 . . . *tournefortiana* Reichb+m
2300-4700 . . . *maritima* L.
2600-3600 . . . *amygdalina* Dcne
2700-4500 . . . *brevifolia* Wall ex DC, *stricta* Edgew.
2750 . . . . . . ? *dolichocephla* Paml*, ? *galuca* Pall ex Willd*
2800-5500 . . . *moorcroftiana* Wall ex DC
3000-5000 . . . *biennis* Willd, *desertorum* Spreng, *dracunculus* L.+m *rutifolia* Steph ex Spreng.
-5500 . . . *sacrorum* Ldb
3400-5500 . . . *macrocephala* Jacq ex Bess.
4000-(plains to 4000) . . *scoparia* Waldst et Kit+m
-5800 . . . *minor* Jacq ex Bess.
5000-9000 . . . *aucheri* Boiss, *campbellii* Hk et T ex CB Cl, *dubia* Wall, *salsoloides* Willd var *paniculata* HK f, *eriocephala* Pamp*.

---

* species dubia (RRS)
† Kazakhistan
\+ Uzbekistan

Central Asia

(Species for which data are available only listed)

Badachschan		Pamir
1700-1800m	*glandulifera*	
-1900	*porrecta*	
-3150	*vulgaris*	
-3200	*scoparia*	
-3500	*dracunculus*	
1800-2800	*tournifortiana*	
-3100	*ferganensis*	
-3300	*vachanica*	
1900-4500	*persica*	3600-3900
2000-3250	*sieversiana*	
2600-4400	*santolinifolia*	3600-4200
2800-3600	*karshinskyi*	
2900-3450	*demissa*	3800
-4400	*rutifolia*	3600-4300
-4500	*macrocephala*	3600-4100
3500-4200	*pamirica*	3600-4100
3600-4400	*leucotricha*	3700-4300
4000-4200	*alcockii*	
	rhodantha	3600-4100
	kuschakeviczii	3600-3900
	viridis	3900-4300

Species reaching Pamir: *macrocephala, persica, rutifolia.* Their distribution is as follows:

persica: Kurram, Chitral, Swat, Ladakh, Balti, Dras, Zoji-la, Kagan valley (Hazara).

rutifolia: Kurram, Gilgit, Balti, Dras, Shyok.

macrocephala: Chitral, Nubra, Sasserla, Ladakh, Sakti, Rupshu, Zanskar-Ldk Road, Kanrgil, Zanskar, Kashmir.

Species reaching Badachschan: *vulgaris, scoparia, tournifortiana, persica, sieversiana, macrocephala, rutifolia.* Their distribution pattern is as follows:

scoparia: Sind, Baluchistan, Kurram, Peshawar, Lahore, Chitral, Swat, Gilgit, Astor, Leh, Zanskar, Kashmir.

sieversiana: Chitral, Balti, Nubra, Gilgit, Leh, Zanskar, Dras, Kashmir.

tournifortiana: Baluchistan, Kurram, Chitral, Swat, Balti, Nubra, Gilgit, Lamayurn, Zanskar, Hazara, Muree, Kashmir.

Aster

In Kashmir we have 12 species: *flaccidus* Bge is a native of Siberia. Five species are endemic; *falconeri* (CI) Hutch (Kash-Pak), *steroides* (DC)Ktze. *idamellus* Grisers (W Him), *peduncularis* Wall ex Nees and *thomsonii* Cl (W.C Him).

Common

Uzb-Kg . . *novae-angliace* L., *vvedenskyi* Bond.

Not common

Uzb . . *papovii* Botsch, *salignus* Willd.

Kg . . *nova-belgii* L, *telmatschevii* Tamamasch.

Pam . . *alpinus* auct, *serpentimontanus* Tamamsch.

Karak . . *flaccidus* Bge.

Kash . . *milliusculus* (DC)CB Cl, *flaccidus* ssp *flaccidus* Grierson.

Cirsium

In Kashmir seven species occur: *falconeri* (Hkf) Petrak is endemic (Kash-Pak), *arvense* is circumpolar, *segetum* Bge is a native of China.

Common

Uzb-Kg-Pam . . *esculentum* (Sievers) CAM, *ochrolepidium* Jaz.

-Kg . . *brevipappisum* Tschern, *glabrifolium* (Winkl)O & B Fedt, *semenovii* Rgl, *sieversii* (F et M) Petrak, *turkestanicum* (Rgl) Petrak.

Not common

Uzb . . *lanceolatum* (L.) Scop.

Kg . . *alatum* (Gmel) Bobr, *alberti* Rgl et Schmath, *sairamense* (Winkl) O & B Fedt, *vulgare* (Savi) Ten.

Pam . . *acaule* T

Karak . . *incanum* (Gmel) F ex BB (Only species!)

Kash . . *arvense*(L.) Scop, *falconeri* (HK f) Petrak, *involucratum* DC, *segetum* Bge, *wallichii* DC.

Jurinea

The genus is rich in Uzb (33 species) and Kg (26 sps) but we have only four taxa. *J. olgae* Rgl et Schmath is common to Uzb-Kg-Pam and *lasiopoda* Tarute to Uzb-Pam, The other two taxa are endemic: *himlaica* RRS var *tibetica* RRS and *ceratocarpa* RRS (Kash-Pak), *ceratocarpa* (Decne) Cl *var-depressa* Cl ex Hk f also occurs in Caucasia, Armenia and Pamir.

Arabidopsis

Five species are reported from Kashmir. Of these *thalliana* (L) Heynh is abundant in Eurasia, *stricta* (Camb)N Busch is endemic (W Him).

Common

Kash-Uzb-Kg . . *thaliana*

-Uzb-Pam . . *pumilla* (St)N Busch

-Kg-Pam. . . *mollissima* (CAM)OES.

Not common

Uzb . . *parvula* (Schrenk)OES.

A. pumila and *A. thaliana* reach Baluchistn. *A. millissima* var *dentata* OES is the only taxon in Karakoram.

Arabis

Nine species occur in Kashmir : *alticola* OES (Kash) and *saxicola* Edgew (Kash-NW Pak) are endemic.

Common

Turk-Uzb-Kg-Kash . . *auriculata* Lam

-Kg . . .*bucharica* (Lipsky) Neveski, *caucasica* Willd.

-Uzb ... *montbretiana* Boiss.

Uzb-Kg Pam . . . *kokanica* Rgl.

-Pam . . . *taraxaifolia* Anders

Pam-Kash . . . . *tibetica* HK f & T (only species in Karakoram)

Not common

Turk . . *ionocalyx* Boiss.

Uzb . . . *karategina* Lipsky, *papovii* Botsch et Vved.

Kash . . . *alticola* OES, *amplexicaulis* Edgew, *bijuga* Watt, *pangiensis* Wall., *pterosperma* Edgew, *pterosperma* var *ramosa* Nbl, *saxicola* Edgew, *tenuirostris* OES.

Saussurea

In Kashmir 31 species occur: *gossypiphora* D Don reaches W China, 18 species are endemic:

Kash . . *candolleana* Cl.

Ladakh . . *dainellii* Pamp, *glanduligera* Sch-Bip, *subalata* Cl, *thomsonii* Cl.

Kash-Pak . . *atkinsonii* Cl. *lappa* (Decne) Sch-Bip, *obvallata* (DC) Sch-Bib

-Ladkh . . .*graminifolia* Wall ex Hk f. *lenotodontoides* (DC) Lip-Sch, *sacra* Edgew.

Ladakh-Pak . . *bracteata* Decne, *nana* Pamp, *stolizkae* Cl

-Kash-Pak . . *falconeri* Hk f

W Him . . *roylei* (DC)Sch-Bip, *jacea* (Koltz)Cl, schultzii Hk f, *taraxacifolia* Wall ex DC.

Four species occur in Uzb: *chondrilloides* C Winkl, *elegans* Ldb, *salsa* (Pall) Spreng, *salosa* Lipsky; 23 species in Kg. *S elegans* and *salsa* are common to Uzb!

Braya

Of the five species in Kashmir *trinkleri* Schmidt is endemic(Ladakh).

Common

Kash-Kg-Pam . . *oxycarpa* HK f & T (reaches W China), *rosea* Bge (circumpolar).

Not common

Kg . . . .*scharoshorsti* Rgl et Schmath

Pam . . *brachycarpa* Vass

Kash . . *aenea* Bge (= *resea* Bge var *albiflora* Maxim). *tibetica* HK f & T (reaches W China)

Draba

Eighteen species occur in Kashmir: *fladnitzensis* Wulf has wide distribution extending to Europe. The endemics are: *aubrietioides* Jafri, *nubigena OES, winterbottomii* (HK f & T) Pohle (Kash); *ludlowiana* Jafri, *tenerrima* OES (Kash-Ladakh), *cachemirica* Gaud, *falconeri* (Cl) OES (Kash-Pak).

Common

Turk-Uzb-Pam-Karak-Kash. .*stenocarpa HK f & T*

-Uzb-Kg-Kash. . lanceolata Royle, numerosa L.

-Uzb-Kg-Pam. . *lanceolata.*

-Uzb-Kg . . . .*alberti* Rgl et Schmath, *huetii* Boiss, *melanopus* Kom, *numerosa* L.

Kg-Pam-Kash . . .*altaica* (CAM)Bge, *lasiophylla* Royle (also in Karakorm), *tibetica* HK f & T.

Pam-Karak-Kash . .*oreades* Schrenk

Not common

Kg . . . *arseniewi* (B Fedt)Gilg, *aucheri* Boiss, *fedtschenkoi* (Pohle)Gilg, *fladnizensis* Wulf, *lipskyi* Tolm, *parviflora* Schulz, *physocarpa* Kom, *sibirica* (Pall)Thellung, subamplexicaule CAM, *telassica* Pohle, *turczaninovi* Pohle et N Busch.

Pam . . . *korschinskii* (O Fedt)Pohle, *pamirica* (O Fedt)Pohle.

Kash . . . *afghanica* Boiss ex Busser, *glomerata* Royle, *glomerata* var *dasycarpa* OES, *setosa* Royle, *trinervis* OES.

Megacarpaea

Two species are reported from Kashmir, both are endemic: *bifida* Bth (Kash) and *polyandra* Bth (Ladakh). These alpines are palatable to browsing animals. The genus is not reported from Pamir.

Common

Turk-Uzb-Kg . . *gigantea* Rgl (only species in Turk!)

Uzb-Kg. . . . . *Orbiculata* B Fedt

Not common

Uzb . . . *megalocarpa* (Fisch)Schisk

Kg . . . *gracilis* Lipsky

Carex (Fig. 62)

In Kashmir 61 species occur, Of these 12 are endemic:

Kash-Ladakh . . *haematostoma* Nees var *submacrogyne* Kuk, *stenophylla* Wahl var *longipedicellata* (Boeck) Kuk.

Ladakh . . *oligocarpa* Cl.

Kash . . . *curta* Good, *flutea* RRS, *kashmirensis*

Kash-Ladakh-Pak . . *lutobrunnea* (Kuk)Nelmes, *plectobasis* Krecz.

-Pak .... *polyphylla* K et K,*vulpinaris* Nees ex Wight forma *angustifolia* Kuk,

W Him .... *pseudobicolor* Boeck

Common

Uzb-Kg-Pam-Kash . . *orbicularis* Boott, *pseudofoetida* Kuk ex Osten, *soongorica* K et K.

-Kg-Pam .... *melanthus* CAM, *oxyleuca* Krcz, *pamiriensis* Cl, *pseudofoetida* Kuk, *stenocarpa* Turcz, *stenophylloides* Krcz, *microglochin* Willd.

-Kg-Kash . . . *acutiformis* Ehrb, *diandra* Schrenk, (var *major* Koch Reymond in Kash), *philocrena* V Krcz, *polyphylla*

-Kg .... *arcatica* Meinsb, *canescens* L, *coarcta* Booth, *compacta* Lam. *decaulescens* V Krcz, *diluta* MB, *duriusculiformis* Krcz, *karoi* Freyn, *knorringiae* Kuk, *koshewnikowii* Litw, *litnowii* Kuk, *melaniostachya* MB, *pachystylis* Cav, *panicea* L. *parva* Nees, *regeliana* Kuk, *riparia* Curt, *turkestanica* Rgl.

Pam-Kash .... *oliverii* Bocklr.

Not common

Kash . . . *brunnea* Thunb, *cariolepis* Nees, *cernua* Boott, *cruenta* Nees, *curta* Gooden, *curta*var *maxima* (Kuk) RRS, *major* (Koch) Reymonds, *divisa* Hunds, *coarcta* Boott, T Koyama, *duthei* Cl. *fedia* Nees, *filicine* Nees var *meiogyne* (Nees) Strachyey foliosa D Don, *gilesii* Nelmes, *haematostom* Nees *inanis* Kunth, *inflata* Huds, *infusa* Nees, *kashmirensis* Nees, *longicruris* Nees, *luteobrunnea* (Kuk) Nelmes, *maritima* Gunner, *malenantha* CAM, *melantantha* var *gracilescens* Kuk *microglochin* Whal, moorcroftiana Falc ex Boott, *myosurus* Nees var *eminens* (Nees) Cl. *nigerrima* Nelmes, *nubigena* D Don, *obscura* Nees, *obscura* var *brachycarpa* Cl, *setigera* D Don var *schlgintaweitiana* (Boeck)Kuk, *setosa* Boott, *stenophylla* Wall, *stenophylla* var *longipedicellata* (Boeck) Kuk, *vesicaria* L. *vulpinaris* Nees, *vulpinaris* forma *angustifolia* Kuk.

Dipsacus

Only two species occur here: *strictus* D Don is endemic (W,C Him) and *nitis* D Don reaches W China and E Him.

Common

Turk-Uzb-Kg . . *laciniatus* L (=*azureus* Schrenk).

Not common

Kg . . .*dipsacoides* (K et K) Botsch, *sativivius* Henck.

Turk . .*strigosus* Willd.

Corydalis (Fig.63)

ca 25 species occur in Kashmir, none are common to Europe, 10 are endemic:

Kash . . *cashmeriana* Royle, ? *crassifolia* Royle, hendersonii Hemsl

Kash-Pak. .*crassissima* Camb, *crithmifolia* Royle, *falconeri* HK f & T, *onobrychis* Fedds, *stewartii* Fedde *thyrsifolia* Prain, *tibetica* HK f & T.

Common

Turk-Uzb . . *papovii* MP

-Uzb-Kg . . *glauscens* Rgl, *gortschakovii* Schrenk, *ladebouriana* K et K, *paniculigera* Rgl et Schmath.

-Kg-Pam . . *gortschakovii*

-Kash . . . *scheloznovianum* Rgl et Schmath

Pam-Kg .... .*fedtschkoana* Rgl, *stricta* Steph

-Kash . . *crassifolia* Royle

Not Common

Turk . . *aitchisonii* MP, *cheionophila* Czern, macrocaulyx Litv.

Uzb . . *bucharica* MP, *darvasica* Rgl, *nevskii* MP, *nudicaulis* Rgl, *pseudostricta* MP, *schangtnii* (Pall)B Fetd, *severtzovii* Rgl.

Kg . . . *akschijrak* NIK, *capnoides* (L.)Pers. *kashgarica* Rgl, *nikitinae* Ait, *pseudoadunca* MP, *semenovii* Rgl.

Pam . . . *fimbrilifera* Korsch, *tenella* K et K.

Kash . . adiantifera Korsch, HK & T, *alpestris* CAM, *Clarkei* Prain, *cyrtocentra* Prain, *diphylla* Wall, *flabellata* Edgew, *govaniana Wall ex Tent, longipes* DC, *ramosa* Wall ex HK & T, *ramosa* var vaginans HK & T.

Karak . . *moorcroftiana* Wall (the only species!).

Gentiana (Figs. 29,63)

In Kashmir 35 species are reported, 14 are endemic:

Kash . . *clarkei* Kusnez var *acuminata* (Cl)Kusnex.

Kash-Ladakh . . *aquatica* L var *karelinii* Cl, *carinata* Griseb var *intermedia* Cl, *huegelii* Griseb, *moorcroftiana* var *maddenii* Cl, *thomsonii* Cl.

Kash-Pak . . *cachemirica* Decne, *venusta* Wall ex Griseb, *burkillii* H Sim (also in Ladakh)

W Him . . . borealis Bge, *moorcroftiana* var *falconeri* Cl, *carinata* Royle, *kuroo*

W,C Him . . . *capitata* Buch-Ham ex D Don.

Common

Turk-Uzb-Kg. . . *olivieri* Grieseb, (only species in Turk!)

-Uzb-Kg-Pam . . *karalinii* Griseb, *leucomelana* Maxim

-Kg . . . . *riparia* K et K, *olgae* Rgl et Schmath.

Kg-Karak-Kash. *tianashanica* Rupr ex Kusn

-Pam-Karak . . *falcata* Turcz

-Pam . . . . . *azurea* Bge, (also in Kash), *pygmaea* Rgl et Schmath, *vvednskyi* Grosch.

Kg-Kash . . *squarrosa* Ldb.

Not common

Kg . . *algidu* Pall, *kaufmanniana* Maxim, *uniflora* Georgi

Pam . . *turkestanorum* Gaud (also in UZB).

Karak . .*borealis* Bge, *coronata* Royle, *marginata* Griseb, *prostrata* Haenke, *prostrata* var karelinii Griseb, *pseudoaquatica* Kusnez, *serrata* Gunn.

Kash . .*aquatica* L, *argentea* (Royle ex D Don)DC, *clarkei* Kusn, *humulis* Stev, *leucomelaena Maxim, marginata* (D Don) Griseb, *moorcroftiana* Wall ex G Don, *pedicellata* (D Don)Griseb, *pedunculata* Royle ex G Don, *prostrata* Haenk, *stocliczkai* Cl, *stracheyi* (Cl)Kitam, *stricta* Klotz, *umbellata* MB var *holosteoides* (Schott et Klotz)Cl.

Gentianella

G. pseudoplumonaris H Smith is the only species in Kashmir, endemic to Ladakh; *azurea* (Bge)Holub is common to Kg and Pam, Gilgit and Ladakh; *turkestanorum* (Gaud)Holub occurs in Uzb only.

Lomatogonium (Fig. 30)

Four species occur in Kashmir: *spathudatum* (Kern) Fernald is endemic (Kash). Other species range as follows:

carinthianum (Wulf) Reichb. . N Amer, Eur, Kamchatica, Jap, C & E Alps, Carpathians, Caucasus, China.

rotatum (L.) Friesex Nym. Arctic Eur - N Amer-Jap.

thomsonii (Cl)Fern . . restricted - Pamir.

Swertia

In Kashmir 12 species occur, *marginata* Schrenk reaches Siberia, eight are endemic:

Kash . . *speciosa* D Don

Kash-Pak . . . *ciliata* (G Don)Burtt, *petiolata* D Don *petiolata* var *lahuensis* (A Kerner)Cl, *speciosa* var *perfoliata* (G Don)Cl, *speciosa* Royle.

W Him . . *chirayita* Rehb ex Fleming) *Karst*, *tetragona* (Edgew)Cl

Common

Uzb-Kg . .*gracilliflora* Gontsch, *gonczaroviana* Pissjauk (=*erosula* Gontsch).

Kg-Kash . .marginata Schrenk

Not common

Kg . . *connata* Schenk.

Kash . .*alata* (D Don)Cl, *angustifolia* Ham ex D Don, *cordata* (G Don)Cl, *paniculata* Wall, *thomsonii* Cl.

Geranium

In Kashmir 18 species are reported, two are endemic: *kishtwariensis* A Knuth and *aconitifolium* sensu Edgew (Kash-Pak).

Common

Khz-Uzb . . *collinum* Steph, *divaricatum, pusillum, robertianu, rotundifolium, sibiricum, transversale.*

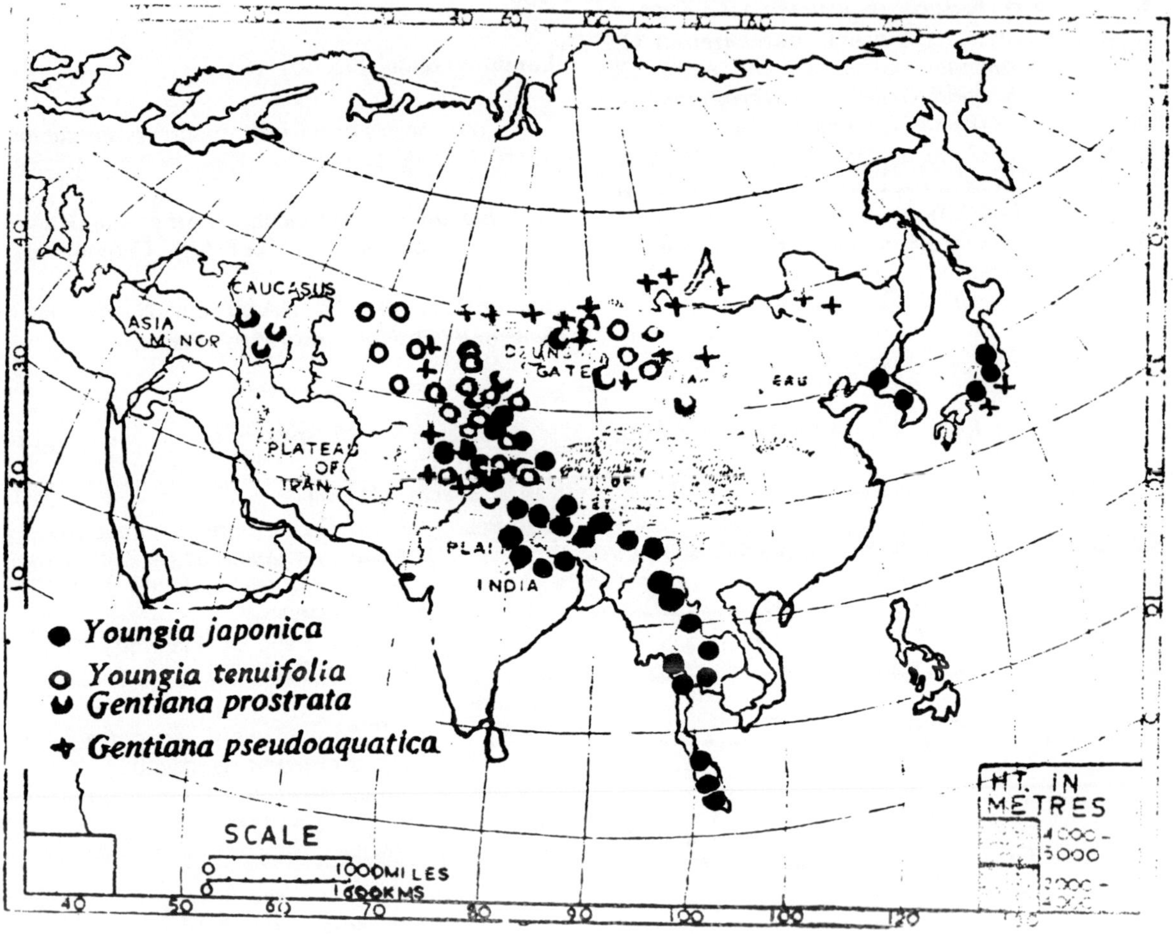

Fig. 29 : *Youngia* is a temperate genus, *japonica* extends into SE Asia and Japan, *tenuifolia* into Siberia, both do not occur in C and mainland China. Of our gentianas only one extends into Japan and one into Caucasus. The genus occurs in N and S temperate regions. A number of species are endemic.

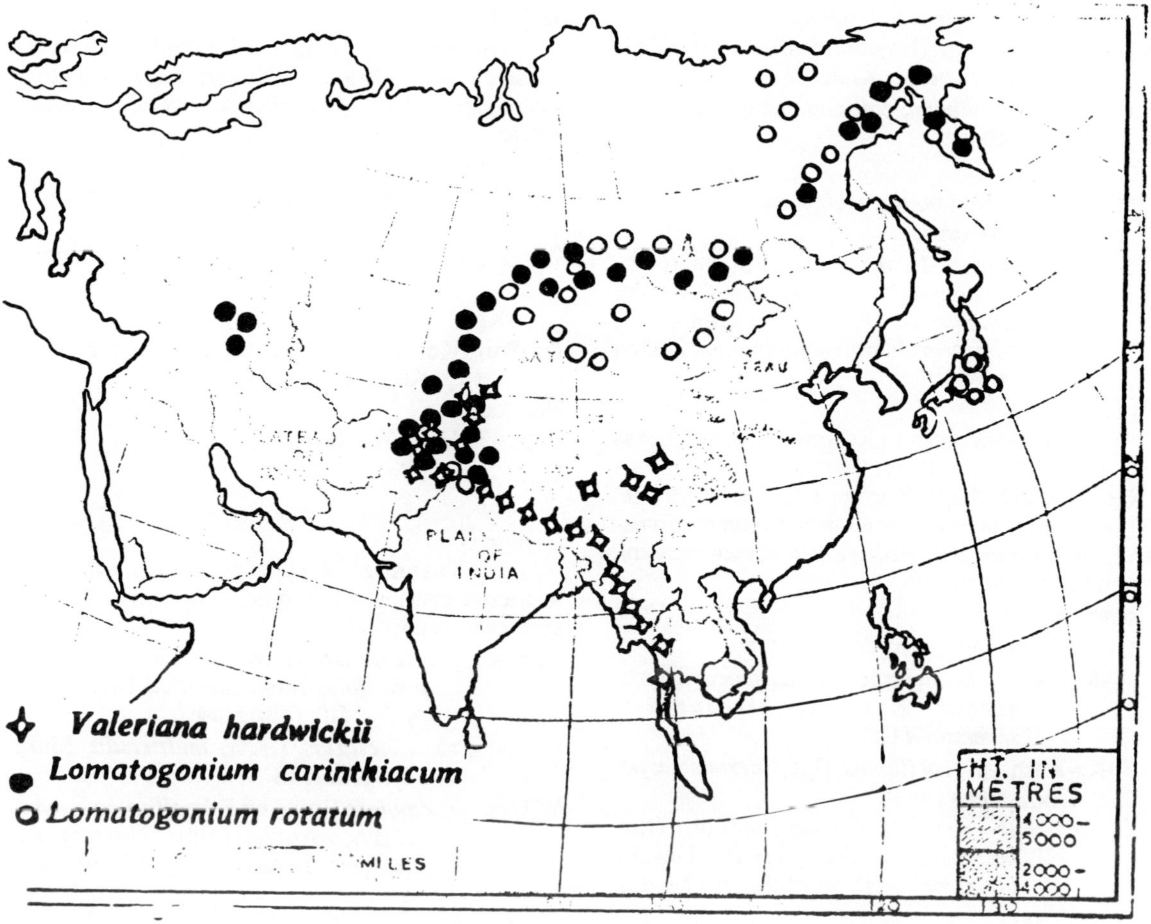

Fig. 30 : *Lomategonium* is yet another taxon which bounces over Japano-Himalayan region; while *rotatum* reaches Japan; *carinthiacum* stops at Korea but extends into Caucasus and the surrounding mts. The genus otherwise is widespread and reaches N America.

-Pam .. *regelii* Nevski, *ferganense* Babr.
-Uzb-Pam-Kash .. *collinum, pusillum* Burm f, *robertianum* L, *rotundifolium*
-Uzb-Kash .. *sibiricum* L.
-Uzb-Pam .. *transversale* (K et K)Vved
-Pam ...... *saxatile* K et K
- Kg ...... *ferganense, regelii*
Kg-Pam . . . albiflorum, sophiae B Fedt, *himalayense* Klotzsch.
Kash-Karak .. *collinum, collinum* var *glandulosum* Ldb

Not common
Uzb . . . *dissectum, linearolobum.*
Pam . . . *tuberosum*
Kash . . . *lucidum* L, *meeboldi* Briq, *molle* L, nepalensis Sweet, *ocellatum* CAM var *himalaicum* R Kn, *pratense* L, *tuberaria* Camb, *wallichianum* D Don ex Sweet

Bromus

Ten species occur in Kashmir ; *inermis* Leyss var *villosus* Mert et Koch (Kash) and var *confinis* (Nees et Steud) Stapf (Kurram-Kash) are endemic. B. *inermis* is the only species in Karakoram, *arvensis* L, *gracillimus* Bge, *mollis* L and *oxydon* Schrenk extend into Baluchistan.

Common
Uzb-Kg-Pam-Kash .. *oxydon* Schrenk
-Kg-Pam ... *denthoniae* Trin, *japonicus* Thunb, *papovii* Drab, *squarrosus* L, *severtzovii* Rgl.
-Kg-Kash . . *gracillimus* Bge, *inermis* Leyss, *scoparium* L.
-Kg . . *angrenicus* Drab, *benekene* (Ldb) Trim, *macrostachys* Desf, *sterilis* Nevski, *turkestanicus* Drab, *tetthelepis* Nevski.

Not common
Uzb . . . *sericeus* Trob, *tytthanthus* Nevski.
Kash . . *arvensis* L, *mollis* L.

Phleum

Three species are recorded here: *himalaicum* Miq is endemic (Chitral-Poonch).

Common
Kg-Pam-Kash . . *paniculatum* Huds
-Pam .... *phleoides* (L.)Simk
Uzb-Kg-Karak-Kash . . *alpinum* L.

Not common
Uzb . . *roshevitzii* N O Pavl
Kg . . . *pratense* L.
Pam . . *graecum*
Karak . . *australis* (Cav) Trin et Steud.

Melica

Of the five species in Kashmir, *nutans* L (Kash), *onoei* Franch (Kagan-Kashtwar)and *scabarrima* (Nees)Hk f (Hazara, Poonch) are endemic.

Common
Uzb-Kg-Pam-Kash . . *jacquemontii* Decne, *secunda* Rgl.
-Kg-Kash .... *inaequeglumis* Boiss, *canascens* (Rgl)Lavr.
-Kg .... *altissima* L.
Kg-Pam-Karak . . *canascens* (only species in Karak!).

Not common
Uzb . . *hohenackeri* Boiss.
Kg .... *transilvanica* Schur.

Nepeta (Fig. 63)

In Kashmir 28 species occur, all have restricted distribution and none reach Europe. 13 species are endemic.

Ladakh . . *cerulescens* Maxim, *nivalus* Bth, *paucifolia* Muk, *thibetica* Bth, *ariostachya* Bth (also Kash).
Ladakh-Pak . . *clarkei* HK f, *multicaulis* Muk, *royleana* RRS
W Him . . . *connata* Royle ex Bth, *elliptica* Royle ex Bth, *govaniana* Bth. (None of our elements are common to E.Him).

Common
Uzb-Kg-Pam . . *formosa* Kudr.
-Kg . . *alatavica* Lipsky, *bracteata* Rgl et Winkl. *kokanica* Rgl, *lipskyi* Kudr, *mariae* Rgl, *manssarifi* Lipsky, *olgae* Rgl, *pannonica* L, *pungens* (Bge)Bth, *sautoana* MP, *satureioides* Boiss, *subhastata* Rgl, *ucranica* L.
Uzb-Kg-Kash . . *cataria* L.
Kg-Pam .... *podistachys* Bth, *ucranica*
Pam-Kash . . . *longibracteata* Bth, *pamierensis* Franch

Not common
Uzb . . *micrantha* Bge
Kg . . . *pseudokokanica* Pojark, *pulchella* Pijark.
Pam . . *daenensis* Boiss, *spatulifera* Bth, *vakhanica* Pojark
Kash . . *botryoides* Solander, *brachyanthus* Rech f et Edelb, *campestris* Bth, *clarkei* Hk f, *coerulescens* Maxim, *connta* Royle ex Bth, *alata* Royle ex Bth, *elliptica* Royle ex Bth, *erecta* (Bth) Bth, *eriostanchya* Bth, *floccosa* Bth, *glutinosa* var *crenata* RRS, *govaniana* (Bth) Bth, *laevigata* (D Don) Bth, *linearis* Royle ex Bth, *longibracteata* Bth, *multicaulis* Muk, *nervosa* Royle ex Bth, *nivalis* Bth, *paucifolia* Muk, *prainii* Duthie, *raphanorhiza* Bth, *royleana* RRS, *tibetica* Bth.

Astraglus

Forty seven species occur in Kashmir, 25 are endemic†. The altitudinal range is interesting in Kash-Pak.

1333-2333 m . . *leucocephalus* Grab ex Bth.
-2550 . . .*trichocarpus* Grab ex Bth.
-3333 . . *amherstianus* Bth ex Royle, *grahamianus* Royle ex Bth.
-4000 . . *graveolens* Buch-Ham ex Bth.
1666- ? . . . . .*altischuri* O Fedt
-3000 . . *bakeri* Ali
2000-3333 . . *chlorstachys* Lindl
-4000 . . *subuliformis* DC
2333-3333 . . *concretus* Bth
-3700 . . *candolleanus* Royle ex Bth
2666-3000 . . *kashmiriensis* Bge
-3666 . . *melacophylla* Bth ex Bge, *falconeri* Bge, *bicuspis* Fisch *chitralensis* Ali.
-4333 . . *maxwellii* Royle ex Bth, *hoffmeisteri* (Kl)Ali ssp *foffmeisteri* Ali.
-4666 . . *himalayanus* Kl
3000-3666 . . *ophiocarpus* Bth ex Bge
-4000 . . *peduncularis* Royle ex Bth
-5000 . . *tribulifolius* Bth ex Bge
3100 . . *breviscapus* Bth ex Bge
3333-4333 . . *marcropteris* Dc, *oplites* Bth ex Parker, *zanskarensis* Bth ex Bge
-4666 . . *frigidus* (L.) A Gray *gracillipes* Bth ex Bge.
3636-5455 . . *polyacanthus* Royle ex Bth[2]
3666-5333 . . *strictus* Grah ex Bth.
4000-5000 . . *melanostachys* Bth ex Bge
-5333 . . *webbianus* Grah ex Bth, *densiflorus* K et K.
4333 . . *pindreensis* (Bth ex Baker) Ali[1], *erionotus* Bth ex Bge[6], *schuganicus* B Fedt[1], *imitensis* Ali
-5500 . . *nivalis* K et K
5000-5455 m . . *hypogottoides* Baker[7]
-5666 . . *confertus* Bth ex Bge, *zacharensis* Bge[3], *coluteocarpus* Boiss var *coluteocadrpus* Ali, *concretus* Bth var *glaber* Ali[5].
5666-6000 . . *arnoldi* Hemsl & Pearson.

Altitudinal range in Central Asia (species for which data are available).

		Badachschan	Pamir
1.	*bakaliensis*	1700-3600 m	
2.	*dsharfi*		-1800
3.	*roschanicus*		-3550
4.	*cysticarpus*		-2600
5.	*colutocarpus*		-3000
6.	*peduncularis*		1800-3400
7.	*tibeticus*		-3500
8.	*olgae*		1900-4100
9.	*xanthomeloides*		-2500
10.	*siversianus*		2000-2400
11.	*orbiculatus*		-2900
12.	*macropteris*	-3600	3800-4000
13.	*stenoceroides*		-3600
14.	*longistipitatus*		2200-2900
15.	*filicaulis*		-3400
16.	*sesamoides*		-2600
17.	*oxypteris*		-2400
18.	*substenoceras*		-2300
19.	*acanthocarpus*		2300-3400

† : endemic in narrow range. 1) Gilgit 2) Ladakh 3) Nubra 4) Kashmir 5) Drass 6) Astor,Seru 7) Pangyong lake and Karakoram.

20.	*aksuensis*		2400-3000
21.	*tribuloides*		-2700
22.	*commixtus*		-3800
23.	*lasiosemius*	-3500	3600-3800
24.	*orczinnikovii*		-2500
25.	*camylorrhynchus*		2500-3000
26.	*alexeenkol*		-3800
27.	*tecti-mundi*	-3500	3800-3900
28.	*breviscapus*	-4100	3700-3800
29.	*schurzae*	2600-3700	3700-3800
30.	*badachshanicus*		2700-3500
31.	*caulescens*		-4000
32.	*nigrocalyx*		2800-3500
33.	*digynous*	-4100	3800-4200
34.	*myriophyllous*	-4000	3800-4200
35.	*vachanicus*		2900-3000
36.	*scheremetewianus*		-3900
37.	*schachaderinus*		2950-4250
38.	*subscaposus*	3000-4050	3800
39.	*bornmuelleranus*		-3100
40.	*schugnanicus*		-4000
41.	*melanostachys*	-4200	4200-4600
42.	*chionanthus*		3100-3500
43.	*peltatus*		3200-4100
44	*alitschuri*		3300-4100
45.	*antoninae*		3450-4150
46.	*dolichopodus*	3500-4000	3700
47.	*chadjanensis*	3600-4000	3600-4100
48.	*pseudobrachytropis*	3800-4200	3800-4200
49.	*charguschanus*	-4500	4200-4600
50.	*orthanthoides*	3900-4900	3700-5000
51.	*nivalis*		4100-4800
	kuschakevitschi		3600-4200
	heterodontus		-4300
	ophiocarpus		3700-3750
	chomutovii		-3900
	alitschuri		-3800
	skorniakovii		3900-4000
	beketovii		-4300
	innominatus		3800-4100
	andaulgensis		3900-4200
	djilgensis		4100-4300
	tulinovii		4000-4550
	tranzschelii		4200-4600

Scutellaria

Six species occur in Kashmir, *prostrata* Jacq ex Bth (W Him) and *teucrifolia* Dunn (Kash-Pak)are endemic; 33 species occur in Uzb and 30 in Kg, both share ten species only.

Not common

Pam . . *leptosiphon* Nevski (also in Turk), *pamirica* Juzz, *paulsenii* Briq (= *oligodonta* Jus) *velutina* Juz et Vved

Turk . . *galericulata* L, *colopoda* Nevski, *nevskii* Juzz et Vved, *squarrosa* Nevski, *luteo-coerulea* Bornm et Sint, *litnovii* Bosum et Sint.

Kash . . *grossa* Wall, *heydei* Hk f, *linearis* Bth.

Caragana

Of the five species reported in Kashmir, *brevifolia* Kom(Ladakh), *vesicolor* Bth (Tibet), *conferta* Bth ex Baker (Kash, Gilgit, Astor, Baltistan) are endemic. There is no record from Pamir.

Common

Uzb-Kg . . *alaica* Pojark, *arborescesns* Lam, *aurantiaca* Koehne, *pleiophylla* (Rgl)Pojark, *turkestanica* Kom.

Not common

Uzb . . *acanthophyllum* Kom, *camilli-schneideri* Kom, *jubata* (Pall) Por, *kirghisorum* Pojark, *laeta* Kom, *laetivirens* Pojark, *leucophloea* Pojak, *psruinosa* Kom, *turfanensis* (Krasch) Kom

Kg . . *grandifolora* (MP)DC

Kash . . *brevispina* Royle ex Bth var *brebispina*, *tragacanthoides* Poir var *himalaica* Kom.

Hedysarum

Six species are reported from here, three are endemic:

Kash-Pak . . *astragaloidea* Bth ex Baker, *cachemerianum* Bth ex Baker.

Kash-Ladakh . . *microcalyx* Baker. H. alpinum L. is circumpolar and *falconeri* Baker renges into Hindukush and W Tibet.

Common

Uzb-Kg . . *alaicum* B Fedt, *angrinicum* Korotk, *omissum* Rgl et Schmath, *flavescens* Rgl et Schmath, *gypsaceum* Kom, *lehmannianum* Bge, *songoricum* Bong.

-Pam . . *poncinstii* Franch

Not common

Pam . . *minjanense* Rech f.

Uzb . . *amukutanicum* B Fedt, *bucharicum* B Fedt, *hemithamnoides* Korotk, *iomuticum* B Fedt, *juxarticum* MP, *kudrjaschavii* Korost, *magnificum* Kudr, *muratense* MP, olgae B Fedt, *papovii* Korotk, *taschkondicum* MP, *vvedenkyi* Korotk.

Kg . . *chaitocarpum* Rgl et Schmath, *denticulatum* Rgl et Schmath, *ferganense* Korsch, kirghisorum B Fedt *krassnovii* Rgl et Schmath, *motanum* (B Fedt)B Fedt neglectum Ldb, *pskemense* MP, *santalaschii* B Fedt, *aschischkinii* Summ, *semenofii* Rgl et Schmath, *setosum* Vved, *sevedrzovii* Bge, *turkestanicum* Rgl et Schmath.

Oxytropis

Ten species occur in Kashmir, four are endemic: *cachemiriana* Camb, *densa* Bth ex Bge, *nivalis* Franch, *parviflora* Franch.

Common

Uzb-Kg-Pam-Kash . . *humifusa* K et K.

-Kg . . *alliovillosa* B Fedt, *glabra* (Lam)DC, *jacunda* Vved, *leucocyana* Bge, *lit vinovii* B Fedt, *pilossima* Vved, *rosea* Bge, *sevellanica* Bge.

-Kg-Pam . . *immersa* (Baker)Bge, *microsphaera* Bge, *pagobia* Bge platonychia Bge.

-Kg-Karak-Kash . . *lappinica* (Wahl)Gray

Pam-Karak-Kash . . *chilliophylla* Royle ex Bth,

Kash-Karak . . *chilliophylla, humifusa, lapponica, mollis.*

The genus is richly represented in Uzb (93 sps) and Kg (50 sps) and 16 species occur in Pamir Alai.

Not common

Pamir . . . *bella* B Fedt, *crassiuscula* Boriss, *gorbunovii* Boriss, *Gentensis* B Fedt, *hedinii* Ulb, *hirsutiuscula* Freyn, *melantricha* Bge, *platyscens* Schrenk, *posinisii* Franch, *tianshanica* Bge, trichosphaera Freyn.

Kash . . . *meinschausenii* Schrenk, *microphylla* (Pall)DC (=*tibetca* Bge), *mollis* Royle ex Bth, *tatarica* Camb. *pusilla* Bge.

Polygonatum

Three species occur here, of these *geminiflorum* Decne is endemic (Kash-Pak), the other two are *multiflorum* (L.) All and *verticillatum* All. None is common to Central Asia. *P.roseum* Kunth Ldb is common between Uzb-Kg-Pam and *P. severtzovii* Rgl between Uzb-Kg.

Herminium (Fig. 54)

Four species are reported: *macrophyllum* (D Don) Dandy is endemic (WH), other species range as follows:

angustifolium (Lindl) Bth ex Cl . .China EHim,. SE Asia.

monorchis (L.)R Br. . Widespread, boreal Asia, Eur

pugioiforme Lindl ex HK f . . E Him.

Androsace

In Kashmir 14 species are reported: *septentrionalis* L (circumpolar) and *villosa* L are widespread; five species are endemic : *aizoon* Duby var, *himalaica* Knuth (Kash-Pak), *muscoides* Duby (Kash), *dutheiana* Kunth and *thomsonii* Wall (Kash-Pak). *rotundifolia* Hardw var *glandulosa* Hk f (W Him).

Common

Turk-Uzb-Pam-Kg.. *maxima* L (=turczaninovii Freyn).

Uzb-Kg . . *sericea* Ovcz, *angrenica* Ivcz.

Kg-Pam . . *acrolasia* Ovcz et Vved, *akbaitalensis* Derg, fedt *schenkoi* Ovcz.

-Karak . *septentrionale* L.

Not common

Uzb . . . *caduca* ovcz, *darvasica* Ovcz, *lehmanniana* Spr.

Kg . . *bungeana* Schischk, *lactiflora* Pall, *olgae* Ovcz, *ovczinikovii* Schischk et Bobr.

Karak . . *dasyphylla* Bge

Kash . . *foliosa* Dcne ex Duby, *lanuginosa* Wall, mucronifolia Wall, *primuloides* Duby, *rotundifolia* Hardw, *sempervivoides* Jacq, *villosa* L.

Primula (Fig. 63)

Of the 21 species in Kashmir, 11 are endemic:

Kash-Pak . . *duthiana* Balf et Sm, *elliptica* Royle, *hazarica* Duthie, *farinosa* L var *armena* C Koch.

Kash . . . . .*reidii* Duthie, *rosea* Royle var *elegans* (Duby) Hk f, *heydei* Wall.

W. Him . . *reptans* Hk F ex Watt, *schilagintweitiana* Pax.

W, C Him . . *minutissima* Jacq ex Duby, *moorcroftiana* Wall ex Kaltt.

Common

Turk-Uzb-Pam . . *fedtschenkoi* Rgl (only in Turk!).

Uzb-Kg . . *algida* Adans, *iljinski* Fedt, *kaulfmanniana* Rgl, *olgae* Rgl, *lectiflora* Turkov.

-Kg-Pam . . *algida*

Kg-Pam . . *pamirica* Fedt, *turkestanica* (Rgl)EA White.

Pam-Kash . . *moorcroftiana* Wall ex Klat.

Kash-Karak .*involucrata* Wall ex Duby, *schalgintwiana.*

Not common

Uzb . . *baldschueanica* B Fedt, *capitellata* Boiss, *geranophylla* Koval, *minkwitzae* WW Sm.

Kg . . .*eugeniae* Fedt, *knorringiana* Fedt, *longiscapa* Ldb, *macroalyx.*

Pam . . *psoralia, aristatus*

Kash . . . *clarkei* Watt, *denticulata* Sm, *erosa* Wall ex Rgl, *floribunda* Wall, *inayatii* Duthie, *macrophylla* D Don, *obtusifolia* Royle, *sibirica* Jacq.

Aconitum

In Kashmir four species are reported: *chasmanthum* Stapf ex Holmes is endemic (Kash-NW Pak), and *soongoricum* reaches W China.

Common

Uzb-Kg . . .*telassicum* Mp, *seravschanicum* Steinb (Only two species in Uzb).

Kg-Kash . . *rotundifolium* K et K, *soongoricum* Stapf.

Not Common

Kg . . *angustecassidatum* Steinb (not in USSR!). *numerum* MP, *septentrionlae* Koelle.

Kash . . *heterophyllum* Wall ex Royle, *violaceum* Stapf, *laeve* Royle.

Anemone

Ten species occur here: *rupicola* Camb var *glabriuscula* Hk f & T is endemic. *rivularis* occurs in Kash-Karak-W China and E Him, *narcissiflora* extends to Europe and *falconeri* to C Asia.

Common

Uzb-Kg . . *protracta* (Ulbr)Juz, *petiolutosa* Juz (also Turk!)

Kg-Pam . . *gortschakowii* K et K.

-Kash . . *obtusiloba* Don

Uzb-Turk . . *tschernjaevii* Rgl, *petiolulosa.*

Not common

Uzb . . *baissunensis* Juz ex Scharipova *bucharica* (Rgl) Fin et Gagnep, *seravaschanica* Kom.

Kg . . . *oligotoma* Juz.

Kash . . *biflora* DC, *falconeri* T T ex Hk, *obtusiloba* var *potentilloides* (Camb)Leueur, *polyanthes* D Don, *rivularis* Buch-Ham, *rupicola* Camb var *labriuscula* Fk HK & T.(*A. protracta* Juz is now *Anemonastrum protractum* (Ulbr) Holub).

Pulsatilla

The only species in Kashmir is *wallichiana* (Royle)Ulbr and is endemic(Kash-N Pak). *campanella* Fisch is common to Uzb-Kg-Pam. In Kg another species is *kostyczewii* (Korsh)Juz.

Thalictrum

Eleven species are reported from Kashmir: *secundum* Edgew (Kash) and *rutaeflolium* HK f & T (Ladakh) are endemic. *T. alpina* L is almost circumpolar.

Common

Uzb-Kg-Kash . . *alpinum* L (also in Karak!)

Turk-Uzb-Kg-Kash . . *minus* L.

-Uzb-Kg . . *isopyroides* CAM, *sultanabadense* Stapf.

Uzb-Kg . . *simplex* L.

Kg-Karak . . *foetidum* L.

Not common

Kg . . *flavum* L, *longipedunculatum* Nik, *petaloideum* L.

Kash . . . *alpinum* var *microphyllum* (Royle) Hind, *chelidonii* DC, *cultractum* Wall, *elegans* Wall ex Royle, *falconeri* Lecoyer, *minus* ssp majus (Crantz)Rony & Four, *pedunculatum* Edgew, *platycarpum* Hk & T, *rutaefolium* Hk & T.

Cotoneaster

In Kashmir 11 species occur: *humilis* Dunn and *obovata* Wall ex Dunn are endemic (Kash-Pak); *acuminata* and *nummularia* occur throughout north temperate region; *uniflora* Bge alone is reported from Karak.

Common

Uzb-Kg-Pam . . *insingis* Pojark, *melanocarpus* Fisch ex Blytt, *oliganthus* Pojakr.

-Pam . . *racemiflora* (Des f)C Koch (=*num mularioides*).

Kg-Pam . . . . . *minutiflora* Bge

-Karak . . . . *uniflora* Bge.

Not common

Kash . . *alatavicus* Mp, *allochrons* Pojark, *karatavicus* Pojark, *krasnovii* Pojark, *neo-antoninae* W Vass, *roborowskii* Pijark.

Uzb . . *soongoricus* (Rgl et Herd) Mp (=*pseudomul tiflora* MP)

Kg . . *suavis* Pojark

Pam . . *discolor* Pojark, *hissaricus* Pojark, *ovatus* Pojark, *splendida* (not in USSR), *umbrosa* (not in USSR)

Kash . . *acuminata* Lindl, *integrerima* Medik, Lindleyi Steud, *microphylla* Wall, *minuta* Klotz ex Lindl, *nummularia* Fisch et Mey, *pruinosa* Klotz, *rosea* Edgew.

Potentilla (Figs. 31, 62)

Potentilla is one of our largest genera. Of the 47 species 13 are endemic:

Kash . . *biflora* Wall ex Schlecht var *armerioides* (HK f)Hand-Mazz, *doubjouneana* Camb, *pteropoda* Royle, *sandersiana* Royle var *jacquemontii* Franch.

Ladakh . . *brevicissa* Bertel, *sericea* var *polyschista* Lehm, *thomsonii* Hand-Mazz.

W and C Him . . *leuchochroa* (Lind)Hk f.

Kash-Pak . . *atrosanguinea* Lodd var *cataclina* Th, *curviseeta* Hk f *monanthes* Lindl ex Lehm.

Four species: *anserina* L, *fruticasa* L, *multifida* L and *nivea* L are circumpolar.

Common

Uzb-Kg-Pam-Kash . . *gelida* CAM, *multifida* L var *angustifolia* Lehm in Kash, *supina* L.

-Kg-Kash . . . *desertorum* Bge

-Kg . . . *approximata* Bge, *asiatica* (T Wulf)Juz, *canascens* Bess, *conferta* Bge, *impolita* Wahl, *orientalis* Juz *soongorica* Bge, *pedata* T Wulf, *tephraleuca* (T Wulf) B Fedt, *reptans*.

-Kg-Pam . . *holoeuca* Boiss ex T Wulf, *evestica* T Wulf, *nervosa* Juz, *pamirialaica* Juz.

Kg-Kash . . . *anserina* Camb, *nivea* L, *biflora* Willd ex Schlecht, *sericea* L, *reptans* L.

-Pam . . . *nudicaulis* Willd ex Schist, *tephroserica* Juz, *virgata* Lehm, *moorcroftiana* Wall ex Lehm.

Not common

Uzb . . . *butkovii* Botsch, *darvasica* Juz ex Botsch. *flabellata* Rgl ex Schmath, *mollissima* Lehm, *tianschanica* T Wulf, *vvdeddenskyi* Botsch.

Galium

Of the 14 species here three are endemic: *ceratophylloides* HK f, *serpylloides* Royle ex HK f, *pauciflorum* Bge (Kash- Pak), *aparine* and *verum* are circumpolar and *boreale* nearly so; *asperifolium* Wall and *elegans* Wall reach W China.

Common

Turk-Uzb-Kg-Pam-Kash . . *spurium* L, *tenuissimum* MB

-Uzb-Kg . . *verticillatum* Danth

-Uzb-Kg-Pam . . *aparine* L (also Kash), *decaisnei* Boiss.

-Uzb-Kg-Kash . . *tricorne* With.

-Kash . . *verum* L.

Uzb-Kg . . *karakulensicum* Pobed, *linczevskii* Pobed, *transcaucasicum* Stepf, *turkestanicum*, *pamiroalaicum*.

-Kg-Pam . . *pamiroalaicum* Pobed.

Kg-Pam . . *saurense* Litw, *septentrionale* Roem et Schult

Not common

Turk . . *coronatum* Sibth et Sow

Uzb . . *nupercreatum* MP

Kg . . *trichophorum* K et K. *vernum* L, puseudorivale Tzvel, *humifusa* Bieb.

Kash . . . *asperifolium* Wall, *asperuloides* Edgew, *boreale* L, *elegans* Wall.

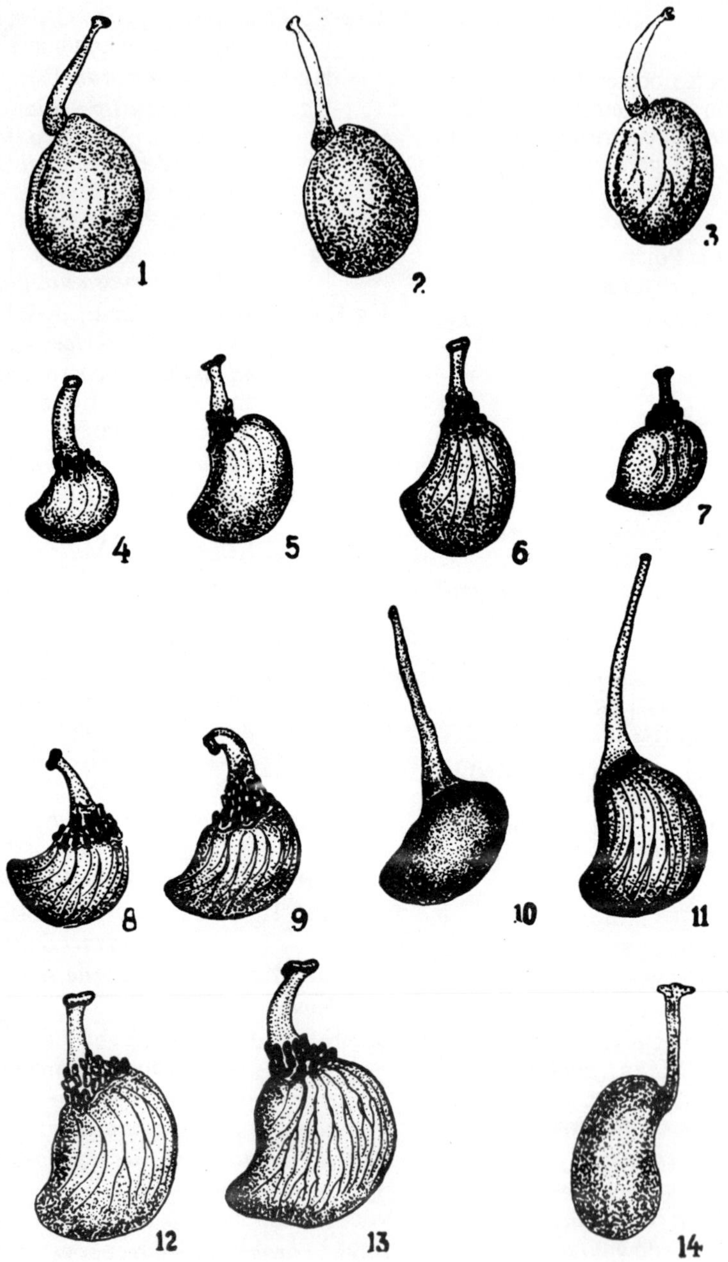

Fig. 31 : Seeds of *potentillas* from Central Asia (courtesy Dushanbe) : (1) *multifida* Linn. ()2) *pamirica* Th Wolf. (3) *schugananica* Juz ex Alyl. (4) *impolita* Wahl. (5) *canascens* Bess. (6) *transcaspia* Th Wolf. (7) *supina* Linn. (8) *arnavatensis* (Th Wolf) Juz. (9) *desertorum* Bage. (10) *flabellata* Rgl ett Schmath. (11) *hijishyana* Th Wolf. (12) *asiatica* (Th Wolf) Juz. (13) *algida* Sojak. (14) *reptans* Linn.

Saxifraga (Fig. 32)

In Kashmir 25 species occur, the endemic are:

Kash-Pak . . *flagellaris* ssp *crassiflagellaris* Hulten, ssp *mumucronulata* (Royle)Engl et Irn, *jacquemontiana* Decane, *lilacina* Duthie, *microphylla* Royle ex HK f & T, *pulvinaria H Sm.*

Kash-Ladakh-Pak . . *stella-aurea* Hk F & T,

Ladakh-Pak . . *meeboldii* Engl et Irn.

W, C Him . . . *brounoniana* Wall ex Sternb, *filicaulis* Wall ex DC, *pseudopallida* Engl et Irm.

Common

Uzb-Kg-Pam-Kash . . *cernua* L, *flagellris* Willd.

-Kg-Kash . . *sibirica* L.

-Kg-Pam . . *hirculus* L, *komarovii* Lose (=*stenophylla* Royle var *hirculus* RRS) *flagellaris* ssp *komarovii* (Loz-Lozin) Hulten.

Kash-Karak . . *hirculus, hirculus* var *alpina* Engl, *hirculus* var *hirculus* RRS, *flagellaris* ssp *kamarovii* (Loz-Lozin)Hulkten.

-Pam . . . *pulvinaria* H Sm.

Not common

Uzb . . . *alberti* Rgl et Schmath, *oppositifolia* L. *setigera* Pursch

Karak . . *androsacea* L var *uniflora* Wulf, *cernua* f *bulbillosa* Engl et Irm, *flagellaris* ssp *euflagellaris* Engl et Irm, *imbricata* Royle, sibirica L var *esuibisrica* Engl et Irm, *op positifolla* ssp *asiatica* (V Hayek)Engl et Irm.

Kash . . ? *afghanica* L, *androsace* var *tridentatus* Gaud ex Meissn, *asarifolia* Sternb, *brunnoniana* Wall ex Sternb, *exigua* H Sm, *filicaulis* Wall ex Ser, *flagellaris* ssp *mucronulata* (Royle) Engl et Irm, ssp *stenophylla* (Royle)Hulten, cf *llonakhensis* WW Sm, *moorcroftiana* (Ser)Wall ex Sternb, *pallida* Wall ex Ser, *pseudopallida* Engl et Irm var *belidofia* Engl et Irm, *ramulosa* E Wall ex DC, *stella-aurea* HK & T, *tangutica* Engl.

Euphrasia

In Kashmir abound 12 species, four are endemic: *foliosa* Penn, *jaeschkei* Wettst, *platyphylla* Penn, *remota* Penn (all Kash).

Common

Uzb-Kg-Pam . . *regellii* Wettst, *tatrica* Fisch (only species in Turk!)

Not common

Kg . . . *bajankolica* Juz, *cyclophylla* Juz, *macrocalyx* Juz, *peduncularis* Juz, *tranzschellii* Juz.

Pam . . *fedtschankoana* Wettst ex Juz.

Kash . . *alba* Penn, *aristulata* Penn, *densiflora* Penn, *falbellata* Penn, *incisa* Penn, *kashmirana* Pugsley, *laxa* Penn, *paucifolia* Wettst, cf *regelii* Wettst, *schlagintweitii* Wettst, cf tatarica Fisch. E. *latifolia* var *flaviflora* Boiss is now *Parentucellia flaviflora* (Boiss) Nevski.(Fl Turk, 6,283, 1954).

Lagotis

Three species occur in Kashmir: *cashmiriana* (Royle)Rupr is confined to Kash-Pak; and *globosa* (Kurz)HK f and *kunawurensis* (Royle)) Rupr endemic to Kash-Pak.

Common

Uzb-Kg . . *karolkovii (Rgl et Schmath) Maz (Only sps in Uzb!)*

Kg-Pam . . *decumbens* Rupr, *ikonnokovii* Schisk (=*cashmirianauct*).

Not common

Kg . . *integrifolia* (Willd)Schischk.

Pam . . *glauca* auct (= ? *L. glauca* Gaertrn, *caschmiriana* auct (in Fl Pam).

Veronica

Of the 26 species eight are endemic here:

Kash . . *leucothrix* Penn.

Ladakh . . *koelzii* Penn, *macrostemon* Bge ex Ldb, *uncinata* Penn.

Kash-Pak . . *cachemirica* Gand, *hirta* Penn, *salina* Schur.

W,C Him . . *secunda* Penn.

Common

Turk-Uzb-Kg-China-Kash . . *anagalis aquatica* L beccabunga L, *arvensisl. persica* Poir.

China-Kash . . *serpyllifolia* L, *biloba* L, *laxa* Bth,

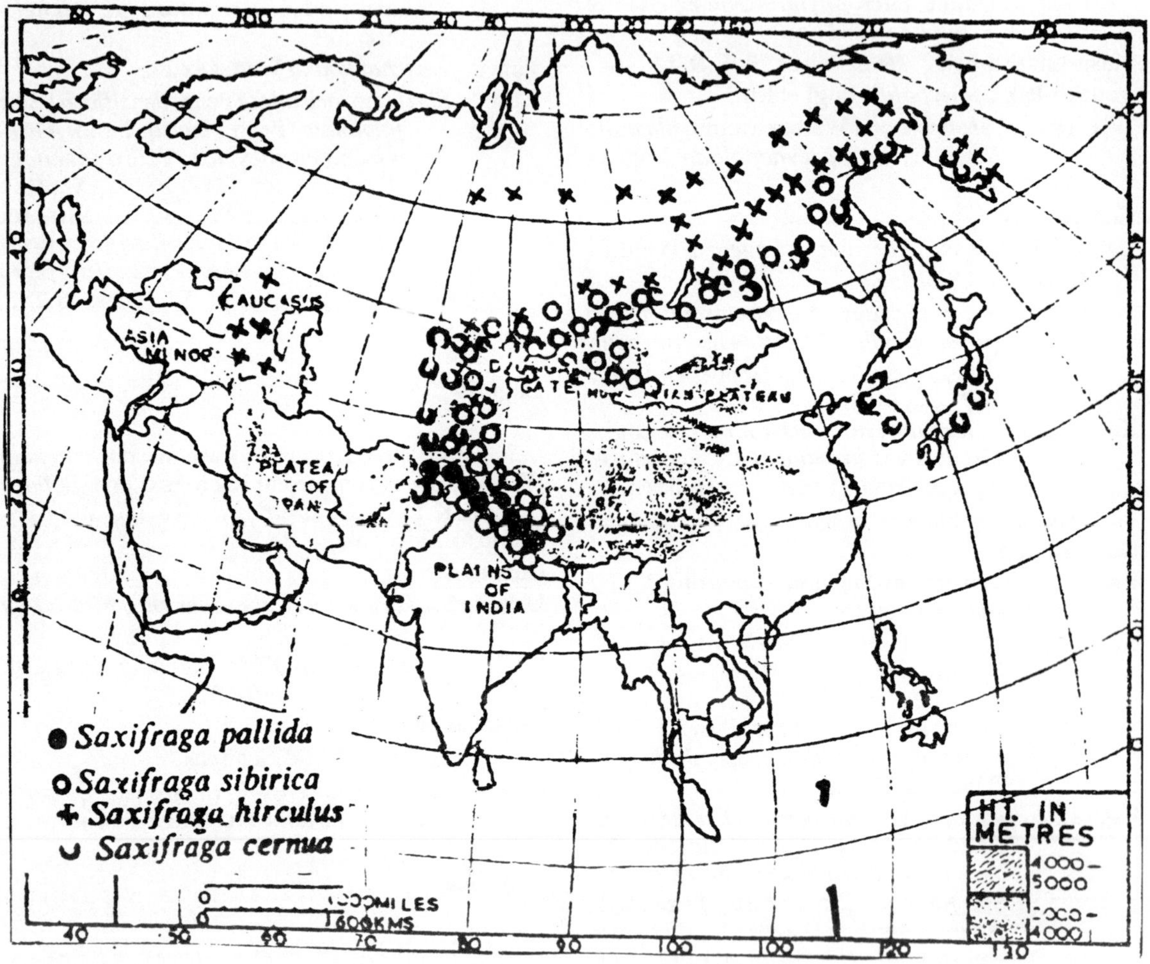

Fig. 32 : *Saxifraga* is widespread, rich in endemics but none of our elements range beyond E Himalaya into SE Asia.

Uzb-Kg . . . . *anagaloides* Guss, *hederifolia* L, *intercedens* Bornm *hispidula* Boiss et Huet (= *purpusilla* Boiss), *rubrifolia* Boiss, *tenuissima* Boriss, tianschanica Lincz, *triphyllos* L.

-Pam-Kash . . *campylopoda* Boiss.

Turk-Kash . . *acampylopoda*, biloba.

-Uzb-Kg-Pam-Kash . . *verna* L.

-Pam . . *michauxii* Lam, *campyolopda*.

Uzb-Kg-Pam . . *hederifolia*, *ruberifolia*.

-Pam . . *cardiocarpa* (K et K) Walp

Kg-Kash . . *serpyllifolia*

Pam-Kash . . *beccabumga*, *biloba* (only sps in Karak!).

Not common

Truk . . *argestris* L, *bobrovii* Nevski, *guadanesis* B Fedt, *gzerniakowskiana* Mouljusch, *kopetdaghensis* B Fedt, *montioides* Boiss, *lysimachioides* Boiss, *khorassanica* Czer, *polita* Fries, *opaca* Fries, *tournifortiana* Gmel, *trukmenorum* B Fedt, *karatavicca* N Pavl, *perpusilla* Nevski.

Uzb . . . *ramosissimum* Boriss, *stylophora* MP, *polita* Fries (also in Kash as didyma Tenn).

Kg . . . *ciliata* Fish, *densiflora* Ldb, *fedtschenkoi* G Boriss, *laeta* K et K, *longifolia* L, *gutkeana* Rupr (not in USSR) *oxycarpa* Bofiss, *porphyriana* Pavl, *spicata* L, *spuria* L.

Pam . . *argute-serrata* Rgl ex Bth, *cardiocarpa* (K et K) Welp, *callipes* (not in USSR).

Kash . . . *cana* Wall ex Bth, *didyma* Tenore, *lanosa* Royle ex Bth, *lasiocarpa* Penn, *lanosa* Royle ex Bth, *macrostemon* Bge ex Ldb, *melissifolia* B Desf ex Poir, *nana* Penn, perpusilla Penn Boiss, *persica* Poir, *stewartti* Penn, *undulata* Wall.

Aulacospermum

Trachydium roylei is endemic (W Him) and the only species here. The genus *Trachydium* is no longer valid; the following comb novo are proposed:

1. **Aulacospermum roylei** (Lindl)Kach, Dhar & Naqshi comb novo (= *Trachydium roylei* Lindl, in Royle Illustr Bot Him 232).
2. **A. dissectum** (C B Cl)Dhar, Kach & Naqshi (= *Trachydium dissectum* C B Cl, FBI 2,672).
3. **A.hirsutulum**(C B Cl) Naqshi, Dhar & Kach (= *Trachydium hirsutulum* C B Cl FBI 2,672).
4. **A.novum-jugum** (C B Cl), Kach, Naqshi & Dhar (= *Trachydium novum- jugum* C B Cl, FBI 2,671).
5. **A. obtusiusculum** (Bth & HK f)Naqshi, Dhar & Kach (=*Trachydium obtusiusculum* Bth & HK f, Gen 2,884; FBI 2,673.)

Ten species of *Trachydium* occur in Uzb, six in Kg and one in Pam.

Common

Uzb-Kg . . *tianschanicum* (Korov)Norm

Kg-Pam . . *simplex* Rupr.

Not common

Uzb . . . *dichotomum* (Korov) Kluykov, *papovii* (Korov)Kluykov.

Kg . . . *froseum* Korov, *tenuisectum* Korov, *turkestanicum* (Franch)Schishk.

Bupleurum

Of the 17 species in Kashmir the endemic are:

Kash . . *jacundum* Kurz var *cachemirica* Cl.

Kash-Pak . . *longicaule* Wall ex DC var *ramosum* Nasir.

W Him . . *aitchisonii* (Boiss)Wolff var *caudatum* Wolff, *longicaule* Wall Dc var *clarkeanum* Wolff, var *maddeni* Cl.

Common

Kash-Pam . . *aitchisonii*

Kash-Uzb . . *linearifolium* DC

-Kg . . . *exaltatum* MB

Uzb-Kg . . . *czimganicum* Lincz

Not common

Kg . . *aureum* Fisch, *dendiflorum* Rupc, *gulczense* O et B Fedt *krylovianum* Schishk, *tianschanicum* Freyn.

Kash . . *candollei* Wall ex DC, *falcatum* L, *falcatum* var *hoffmeisteri* (Kl)Cl, *gracillimum* Kl (also Karak) *lanceolatum* Wall ex DC, *longicaule* var *ramosum* Nasir, *maddeni* Cl, *rotundifolium* L, *subuniflorum* Boiss et Heldr, *tenui* Ham ex Don, thomsonii Cl.

Heracleum

The five species reported from Kashmir range between Afghanistan and W China: *canascens* Lindl, *thomsonii* Cl var *glabior* Cl are endemic (W Him).

Common

Uzb-Kg . . *dissectum* Ldb, *lehmannianum* Bge.

Not common

Kg . . *angustilobatum* Ldb, *candicans* Wall ex Dc, *canascens* Lindl, *pinnatum* Cl, *thomsoni* Cl.

Valeriana (Fig. 30,33)

In Kashmir eight species occur, *dubia* Bge, *hardwickii* Wall var *hoffmeisteri* (Kl)Cl, *pyrolaefolia* Decne and *clarkei* Briq are endemic (all Kash-Pak).

Common

Turk-Uzb-Kg . . *chionophylla* MP ex Kult, ficariifolia Boiss.

Uzb-Kg-Pam . . *feldschnkovii* Coincy

Karak-Kash . . *dioica* L (only sps in Karak!)

Not Common

Kg . . . *officinalis* L, *turkestanica* Summ (= *dubia*Bge).

Turk . . *sisymbrifolia* Desf

Kash . . *hardwickii* Wall, *jatamansi* Jones, *stracheyi* Cl.

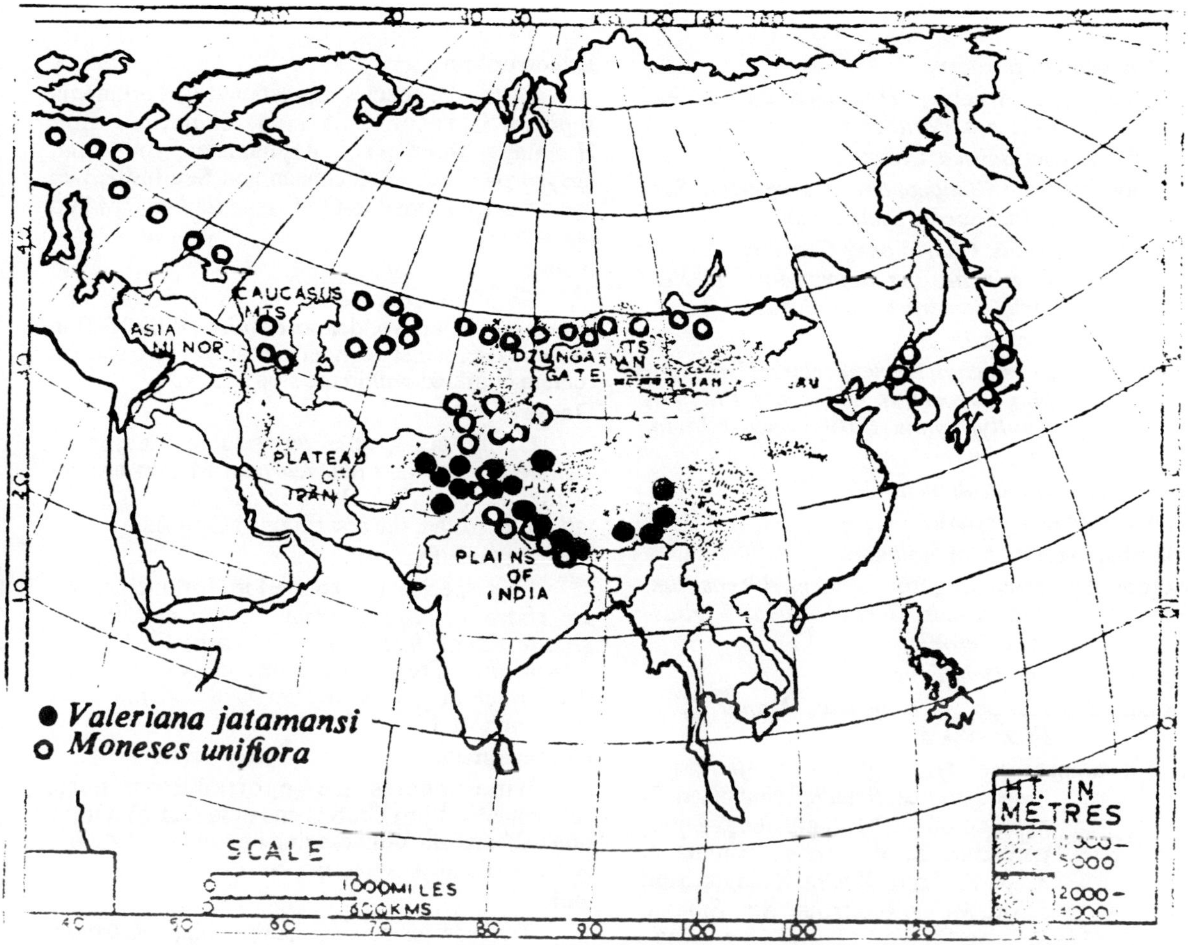

Fig. 33 : *Moneses uniflora*, a boreal arctic element is circumpolar but does not occur in W and C China, *Valeriana* is a large genus occurring in Europe, Asia, Africa and America. *jatamanse* does not range beyond Himalayan arc, but *hardwickii* ranging into SE Asia does not occur in Japan. Our all valerianas are alpine-subalpine.

6

SOME GENERA WITH ENDEMICS IN CENTRAL ASIA

Berberidaceae: *Berberis*

Boraginaceae: *Arnebia, Heliotropium, Nonea, Onosma, Paracaryum.*

Capparidaceae: *Buhsea, Cleome.*

Chenopodiaceae: *Climacoptera, Corisperma, Salsola, Suaeda, Sympegma.*

Cruciferae: *Arabis, Chartoloma, Crambe, Erysimum, Lepidium, Moriera, Peltaria, Sameraria, Torularia.*

Cyperaceae: *Kobresia.*

Labiatae: *Eremostachys, Hymenocrater, Hypogompha, Lagochilus, Origanum, Phlomis, Phyllyphyton, Salvia, Scutellaria, Zizyphus.*

Orobanchaceae: *Orobanche.*

Papaveraceae: *Corydalis.*

Plumbaginaceae: *Acantholimon.*

Scrophulariaceae: *Lagotis, Linaria, Oreosolon, Pedicularis, Scrophularia, Verbascum, Veronica.*

Ranunculaceae: *Delphinium.*

Solanceae: *Atropa, Hyoscyamus, Mandragora, Przewalskia.*

Abbreviations used: *Afg,* Afghanistan, *Afr* Africa, *Amer* America, *Beluch,*Beluchistan,C-M, Central & Median.*Djung,* Djungaria *Eur,* Europe. *Him,* Himalaya. *Kash,* Kashmir. *Kashg,* Kashgar, *Jap,* Japan. *Mong,* Mongolia. *Sib,* Siberia. *Sev,* north. *Vost,* EAST. *zep* West. *iago,* South.

Berberis(Fig. 34)

Only one species occurs and this is endemic: *turcomanica* K et K.

Arnebia

Four species are reported, one *transcapsica* M Pop is endemic: *linearifolia* DC reaches Egypt, the rest are confined to C-M Asia.

Heliotropium

Of the 12 species represented in Turkmenia *supinum* L reaches Africa, *ellipticum,* Ldb Himalaya; three reach Afghanistan, four Iran, *dasycarpum* Ldb, Beluchistan and Sev India; two species are confined to-C-M Asia, *litvinovii* M Pop is endemic.

Nonea

Four species occur in Turkmenia, *turcomanica* M Pop is endemic: *caspica* (Willd)G Don ranges to Beluchistan. India to Malaya. Two species remain confined within the study area.

Onosma

Of the six species reported in Turkmenia, *stamineum* Ldb and *leucocarpum* M Pop are endemic; *dichroanthum* Boiss extends into Afghanistan, Beluchistan, the rest occur in C-M Asia.

Paracaryum (Fig. 36)

Four species are reported in Turkmenia, two are endemic: *turcomanicum* Bornm et Sint and *gracile* Czern; *himalyense* (Klotsch) CB Cl and *intermedium* (Freyn)Lipsky are mainly C-M Asian, the former ranging into Himalaya and the latter Egypt and Arabia.

Climacoptera

Nine species are reported from here; *roborowskii* (Iljin) Grub is endemic (Kash). Of the other species the only ranging beyond C-M Asia is *brachiata* (Pall)Botsch (Eur-Sib).

Buhsea

Two species occur in Turkmenia, *coluteoides* (Boiss, Bge occurs in Iran and *raddeana* (Trautv)Boriss is endemic.

Cleome

Seven species occur in Turkmenia; *turkmena* Bobr is endemic; *chorossancia* Bge reaches Afg, the rest are local.

Corispermum (Fig. 38)

Of the 13 species, two are endemic: *falcatum* Iljin (Caidan, Cindai, Tibet) and *patelliforme* Iljin (Mong, Caidan). Species extending beyond C-M Asia are:

Sib . . *mongolicum* Iljin,

Sib-Eur . . *orientale* Lam.

C. hilaraea Iljin is only reported from Pamir.

Salsola

Three species are endemic: *ikonnikovii* Iljin (Mong), *zaidamica* Iljin (Mong, Kashg, Caidam), and *abrotanoides* Bge (Mongolia, Kashg, Caidam, Djung, Cinai). Of the other 20 species those occurring beyond C-M Asia are:

Eur . . . *praecox* Litw.

Sib . . . *rosacea* L.

Eur.-Sib . . *soda* L, *tamariscina* Pall, *folisa* (L)Schrad.

Eur-Sib-Him . . *collina* Pall, *pestifera* Nels (sev Amer).

Suaeda (Fig. 39)

Seventeen species occur here, only one *przewalskii* Bge is endemic (Mong). Species beyond C-M Asia are the following:

Sib . . *linifolia* Pall, *physophora* Pall, *salsa* (L.)Pall

Eur . . *altissima* (L)Pall.

Sib-Eur . . *acuminata* (CAM)Moq, *corniculata* (CAM)Bge, *prostrata* Pall.

-Korea-Jap . . *glauca* Bge.

Sympegma

S. regelii Bge is the only species and a broadrange endemic to Central Asia (Mong, Kashg, Caidam, Djung, Tienshan).

Arabis

Five species occur in Turkmenia, of these *ionocalyx* (Will) Boiss is endemic: *auriculata* Lam reaches Europe and India, Of the rest lying within C-M Asia, *montbretiana* Boiss reaches Afg.

Chartoloma

Only one species occurs in Turkmenia and is endemic: *platycarpum* Bge.

Crambe

Three species occur in Turkmenia: *gordjagini* Spryg et Pop is endemic, *kotschyana* Boiss and *eduntule* F et M are conifined to C-M Asia.

Erysimum

Seven species occur in Turkmenia, three are endemic: *ischnostylum* Fr et Sint, *gaudanense* Litw and *badghysi* (Korsch)Lipsky; *repandum* L reaches Europe, the rest are local.

Lepidium

Ten species occur in Turkmenia, two are endemic: *kopetdagense* Vved and *kuhitangi* Vass; species ranging beyond C-M Asia are:

Sub-Eur . . *crassifolium* W.K., *perfoliatum* L (India), *ruderle* L.

Eur . . *latifolium* L, *sativum* L, *pinnatifolium* Ldb reaches India.

Moriera

Two species are reported from Turkmenia; *transhyrcana* Czern is endemic, *spinosa* Boiss also occurs in Iran.

Peltaria (Fig. 35)

Only one species occurs in Turkmenia and is endemic: *turkmena* Lipsky.

Sameraria

Three species occur in Turkmenia; *turcomanica* (Korsch)B Fedt is endemic, the other two also occur in Iran-Afghanistan: *aitchisonii* (Korsch) B Fedt and *bullata* (Aitch et Hemsl) B Fedt.

Torularia

Of the five species occurring in Turkmenia, one is endemic: *sumbarensi* (Lipsky)OES; *contortuplicata* (Steph)OES reaches Europe, *adpressa* (Trautv)OES Afghanistan, the other occur in Central Asia.

Kobresia (Fig. 40)

Of the 17 species one is endemic *robusta Maxim (Caidam, Cinhai, Tibet). Species ranging beyond C-M Asia are:*

Eur-Sib-Korea-Jap-Sev Amer . . *bellardii* (All)Degl.

Sib . . . *filifolia* (Turcz)Cl, *stenocarpa* (K et K)Steud, *smirnovii* Ivanova

Eur-Sib-Sev Amer . . *simpliciuscula* (Wahl)Mack.

Him . . *capillifolia* (Kash), *tibetica* Maxim, *pygmaea* (Cl)Cl, *royleana* (Nees)Boeck.

Eremostachys (Fig. 41)

Eleven species occur in Turkmenia, four are endemic: *labiosiformis* (M Pop) Knorr, *boissieriana* Rgl, *bachradenica* B Fedt and *subspicata* M Pop.

Afgh . . *labiosa* Bge, *pesimilis* Aitch et Hemsl, *speciosa* Rupr

Malaya . . *moluccelloides* Bge

Fig. 34 *Berberis turkomanica* Karel.

Fig. 35 *Peltaria turkmena* Lipsky.

37

Fig. 37 : *Hyoscyanus turco manicus* pojark.

36

Fig. 36 : *paracaryum* ***turkomanicum*** Bornm et sint.

Fig. 38 : Coryspermum patelliforme Ijin. Fig. 39 : Suaeda przewalskii Bge. Fig. 40 : Kobresia robusta Maxim.

Hymenocrater(Fig. 43)

Two species are reported in Turkmenia, one is endemic: *incisodentatus* Boriss, the other *bituminosus* Fisch et Mey occurs in Kavkaz and Iran.

Hypogomphia

Only species reported in Turkmenia, *turkestana* Bge, is a broad-range endemic in Central Asia.

Lagochilus

Eight species are reported, two are endemic: *grandiflorus* Wu et Hsuan (Djung) and *lanatonodus* Wu et Hsuan (Kashg, Djung). Of the other six *dicanthophyllous* (Pall)Benth. and *ilicifolius* Bge reach Siberia. In Turkmenia five species occur *balchanicus* Czern is endemic, *nevskii* Knorr also occurs in Pamiro-Alai, and *cabulicus* Bth var *turcomanicus* Bornm et Sint in Afghanistan, Rest lie within C-M Asia.

Origanum

Only one species is reported in Turkmenia and it is endemic: *kopetaghensis* Boriss.

Phlomis

Of the 12 species four are endemic: *admirabilis* Tschern (Mong, Cinhai), *dentosus* Franch (Mong), *kawaguchi* Murata (Tibet) and *similis* Tschern (Mong, Cinhai). Species ranging beyond C-M Asia are:

Eur-Sib . . *tuberosa* L.

Sib . . . . *agraria* Bge, *alpina* Pall, *oreophila* K et K.

Korea . . . *umbrosa* Turcz

Him . . . *younghasbandii* K Mukerjee.

In Turkmenia five species occur, *spinidens* Nevski is endemic, *cancellata* Bge and *kopetdaghensis* Knorr also occur in Iran, and *thapcoides* Bge and *canascens* Rgl in Pamiro-Alai.

Phyllophyton

Five species occur here, *decolorans* (Hemsl)Kudo is endemic (Tibet), all the rest occur in Tibet, *nivale* (Jacq) Cy Wu (zap), *pharicum* (Prain) Kudo (Vost) and *tibeticum* (Jacq)Cy Wu (zap) reaching Himalaya.

Salvia(Fig. 42)

Of the six species, *wardii* Peter-Stibal is endemic (Tibet)), *deserta* Schang reaches Siberia; *nubicola* Wall ex Sweet and *roborowskii* Maxim the Himalaya.

In Turkmenia 12 species occur, four are endemic: *lilacinocoerulea* Nevski, *seravschanica* Rgl et Schmath (sr Asia): and *linczevskii* S Kudr (Kopet-Dagh); *chaoroleuca* Rech f et Aellen does not extend beyond C-M Asia; the other species ranging beyond this area are:

Sv Africa . . *viridis* L(+ Algier),

Arabia . . . *spinosa* L, *sclarea* L.

Eur . . *viridis*, *aethiopsis* L (+ Malaya), *sclarea*.

Afg . . . *bucharica* M Pop, *macrosiphon* Boiss (+ Beluch).

Scutellaria(Fig. 44)

Of the 13 species one is endemic, *alaschanica* Tschern (Mong); species ranging beyond C-M Asia are:

Sib- Korea-Jap . . *baicalensis* Georgi, *scordrifolia* Fisch

Sib- Eur- Jap . . *galericulata* L.

Sib- Eur . . *supina* L.

Sib . . . *grandiflora* Sims, *krylovii* Juz, *sieversii* Bge

Jap . . *viscidula* Bge.

vost Him . . *kingiana*

Eight species occur in Turkmenia, six are endemic: *leptosiphon* Nevski, *colpodea* Nevski, *nevskii* Juz et Vved, *heterotricha* Juz et Vved and *litvinovii* Bornm et Sint. *galericulata* L is very wide-spread (Malaya, Him, Amer sev, cev Afr, Jap, Kitai). *luteocoerulea* Bornm et Sint is local.

Zizyphora

Of the six species in Turkmenia one is endemic: *galinae* Juz, the rest are confined to C-M Asia; *persica* Bge reaches Afghanistan.

Orobanche

Of the 18 species in Turkmenia three are endemic: *borissovae* Novopokr *androssivii* Novopokr, *nikitinii* Novopokr and (?) *inconspicua* Gontsch. *aegyptiaca* Pers reaches Egypt and Him and 12 species are common to Afg and seven to sev Afr.

Corydalis

Four species occur in Turkmenia, *chionophila* Czern and *macrocalyx* Litv are endemic, *papovii* Nevski is local and *aitchisonii* M Pop occur in Iran.

Acantholimon

Of the 15 species in Turkmenia, two are endemic: *korovinii* Czern and *balchaicum* Korov; ten species reach Iran, two (*majevianum* Rgl, *erythraecum* Bge) Pamiro-Alai and one *nikitinii* Lincz Afghanistan.

Fig. 41 : *Eremostachya boissierian* Rgl.

Fig. 42 : *Silvia lilacinocoerulea* Nevski.

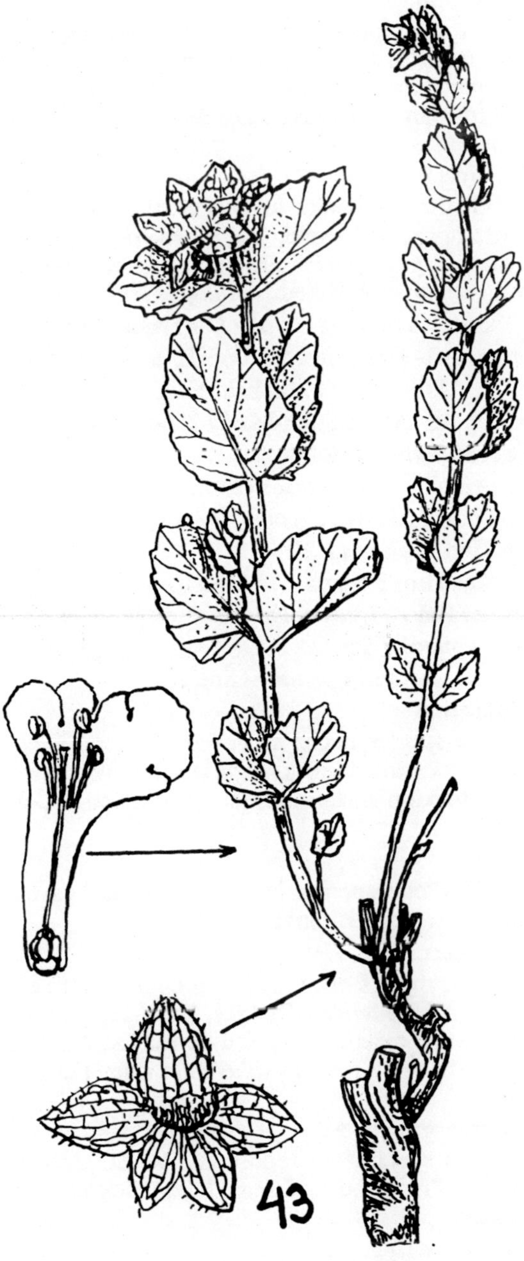

Fig. 43 : *Hymenocrater incisodentatus* Boriss.

Fig. 44 : *Scutellaria litninovii* Bornm et Sint.

Lagotis(Fig. 45)

Of the five species one *brevituba* Maxim is endemic (Cinhai). Species ranging beyond C-M Asia are:

Him . . . *decumbens* Rupr (zap Kash)

Sib . . *integrifolia* (Willd)Schisk.

Linaria

Four species occur in Turkmenia, *striatella* Kupr is endemic, *simplex* DC reaches Arabia and Afr, rest are confined to C-M Asia.

Oreosolon

Only one species is reported and is endemic: *anguiculatus* Hemsl (Tibet).

Pedicularis(Fig. 50)

It is a very large genus with 82 species. Nine are endemic: *chorgossica* Regl et Winkl (Djung), *fetisowii* Rgl (Djung), *junatovii* Ivanina (Djung), *moschata* Maxim (Mong), *pygmaea* Maxim (Tibet) *scolopax* Maxim (Cinhai), *ternata* Maxim (Mong, Cinhai, Tibet), *pseudocurvituba* Tsoong (Caidam, Cinhai, Tibet), *retingensis* Tsoong (Tibet).

Fifteen species are common to Siberia; of these *verticillata* L reaches Jap - Alaska sev Amer; 17 are common to Him (6 to vost) and of the these *semenovii* Rgl and *dolichorrhiza* Schrenk (Kash) occur in Afgh and Kash; and *rhinanthoides* Schrenk and *longiflora Rudolph also in Siberia. Other species range as follows:*

Sib- Korea- Jap . . *resupinata* L, *sceptrum-carolinum* L.

Eur-Sib . . . *amoena* Adams ex Stev, *dasystachys* Schrenk, *physocalyx* Bge, *karoi* Freyn, oederi Vahl (Him, sev Amer).

Scrophularia(Fig. 46-48)

Of the 12 species four are endemic: *alaschanica* Batalin (Mong), *potaninii* Ivanina (Mong), *przewalskii* Batalin (Mong, Cinhai, Tibet), *regelii* Ivanina (Djung). Species ranging beyond C-M Asia are:

Sib . . *alata* Gilib, *altaica* Murr, *huecheriiflora* Schrenk, *incisa* Weinm

Him . . *dentata* Royle ex Bth (zap Kash).

Eleven species occur in Turkmenia,six are endemic: *czerniakowskiana* B Fedt, *nikitinii* Gorschk, *litvinii* B Fedt, *czapandaghi* B Fedt, *turcomanica* Bornm et Sint ex B Fedt: the rest are common to Iran; *xanthoglossa* Boiss extends to Lebanon and Palestine.

Verbascum

Of the species in Turkmenia one, *turcomanicum* Murb is endemic; *thapsus* L is widespread from Europe to Himalaya; six are common to Iran; and *transcapsicum* Murb reaches Malay.

Veronica(Fig. 49)

Of the 24 species in Turkmenia, six are endemic: *bobrovii* Nevski, *czerniakowskiana* Monjuschko, *kopetdaghensis* B Fedt, *turkmenorum* B Fedt, *gaudanensis* B Fedt, *capillipes* Nevski.

Species ranging beyong C-M Asia are:

Eur-sev Afr- Him . . *beccabunga* L, *arvensis* Karak.

Eur- sev Af:- Malay . . *hederifolia* L.

Eur-sev Afr-Afg-Beluch-Malay . . *polita* Fries

Afg . . *montioides* Boiss.

Afg- Arabia . . *angalloides* Guss.

Afg-Beluch-Malaya . . *campylopoda* Boiss.

Afg- Him . . *biloba* L.

Eur- Afg-Him-Malaya . . . *verna* L.

Eur- sev Afr . . *agrestis* L.

Eur . . . *opaca* Fries

Eur-Him . . . *tournifortiana* Gmel.

Delphinium

Six species occur in Turkmenia, one is endemic: *turkmanum* Lipsky. *semibartum* Bienert extends into Afghanistan, the rest are confined to C-M Asia.

Atropa

Only one species in Turkmenia and is endemic: *komarovii* Blinovsky et Schalyt.

Hyoscyamus(Fig. 37)

Four species occur in Turkmenia, two are endemic: *turcomanicus* Pojark and *kopetdaghii* Pojark; *niger* extends from Eur-Jap and also occurs in sev Afr and Amer; *pusillus* L reaches Eur and Egypt.

Mandragora(Fig. 51)

Two species are reported, *tiberica* Grub is endemic (Tibet) and *caulescens* Cl extends into vost Himalaya.

Przewalskia(Fig. 52)

Two species are reported, both are endemic: *shebberei* (CES Fischer) Grub (Tibet) and *tangutica* Maxim (Cinhai, Tibet).

Fig. 45 : *Lagotis brevituba* Maxim. Fig. 46 : *Scrophularia ataschanica* Betalin. Fig. 47 : S. przewalskii betalin. Fig. 48 : *S. regellii* Ivanina.

Fig. 48A *Scrophularia litvinovii* B.fedt. Fig. 49 : *Veronica turkmenorum* B.Fedt. Fig. 50 : *Pedicularis fetisowii*

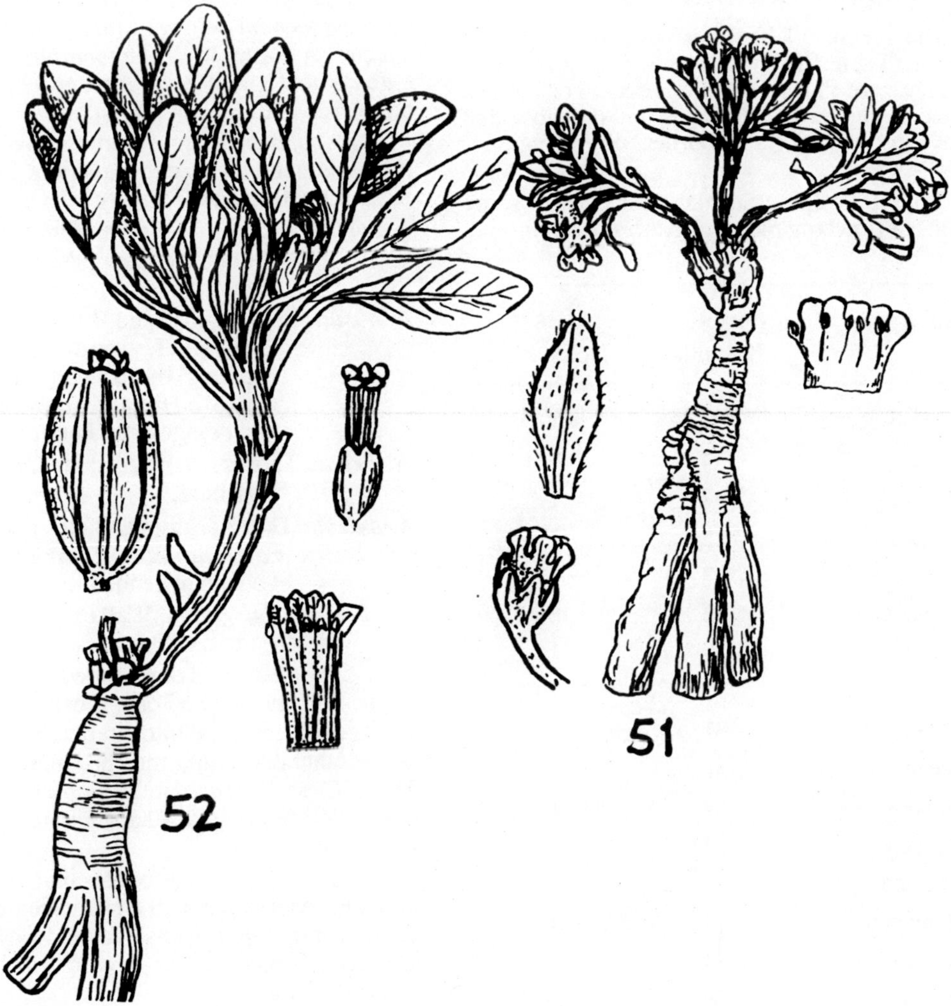

Fig. 51 : *Mandragora tibetica* Grub. Fig. 52 : *Przewalskia tanjutica* Maxim.

7

SINO-JAPANESE-HIMALAYAN PERSPECTIVE

Sino-Japanese elements (Fig. 53)

The species of Kashmir with Sino-Japanese affinities (Table III) show an interesting distribution. These are illustrated by two sets of examples: (a) species that range into Japan and (b) those that 'bounce' over the northern borders of this region.

Table III. Alpine-Subalapine species of some families of flowerng plants of Kashmir common to Japan.

Families	Species in Kashmir	species in Japan
Caryophyllaceae	97	1
Compositae	334	2
Cruciferae	97	1
Cyperaceae	25	3
Elaegnaceae	3	1
Gentianaceae	55	8
Iridaceae	9	1
Juncaceae	15	7
Labiatae	81	1
Liliaceae	46	5
Monotropaceae	4	1
Onagraceae	23	6
Orchidaceae	7	1
Polygonaceae	63	8
Rosaceae	104	9
Saxifragaceae	28	3
Umbelliferae	62	1
Urticaceae	8	1

(a) SPECIES RANGING INTO JAPAN

Hypecoum wide Eurasian genus has two species in Kashmir, *H. leptocarpum* HK. f. & T. extends to Central Asia through (?) Ladakh and eastwards to China and Japan.

Anaphalis though leaning towards Central Asia has wide range and illustrates NW and E of Kashmir extensions (Fig. 28):

A. timmua D Don .. China, E Himalaya, Vietnam, Korea, Japan, Kamachatica(N. America).

A. *busua* (Buch-Ham)Hand.-Mzt. . *A. contorta* (D Don)Hk.f. and *A. nepalensis* (Spreng)Hand.-Mzt., eastwards to China, E Himalaya only; while *A. contorta* ranges into Pamir also.

A. virgata Thoms . . Afghanistan, Iran, Dzungaria-Kaschgar.

A.royleana DC, Kashmir-Ladakh- Pakistan and *A. royleana* var *concolor* HK. f. . Kashmir Ladakh are endemics.

Artemisia roxburghiana Wall ex Bess and *A. dubia* Wall. .. China,Japan, E Himalaya.

Cynoglossum is represented by seven species. Of these C. *lanceolata* Forsk is of special interest as it ranges on one hand into Arabia and N Africa and on the other into China, the Philippines, Taiwan and Japan, C. *zeylanica* (Vahl)Thunb ex Lehm appears to be a southeast Asian element ranging into Japan. (Fig. 27).

Of the five species of *Desmodium*, *D.odocarpum* DC reaches Japan, the other species range into Central and Eastern Himalaya (and western parts of China). However, our Medicagos are wholly Mediterranean !

Of *Elaegnus, E.umbellatus* Thunb is a native of Japan and extends into the Himalaya; while *E. rhamnoides* L. is common in Ladakh and ranges

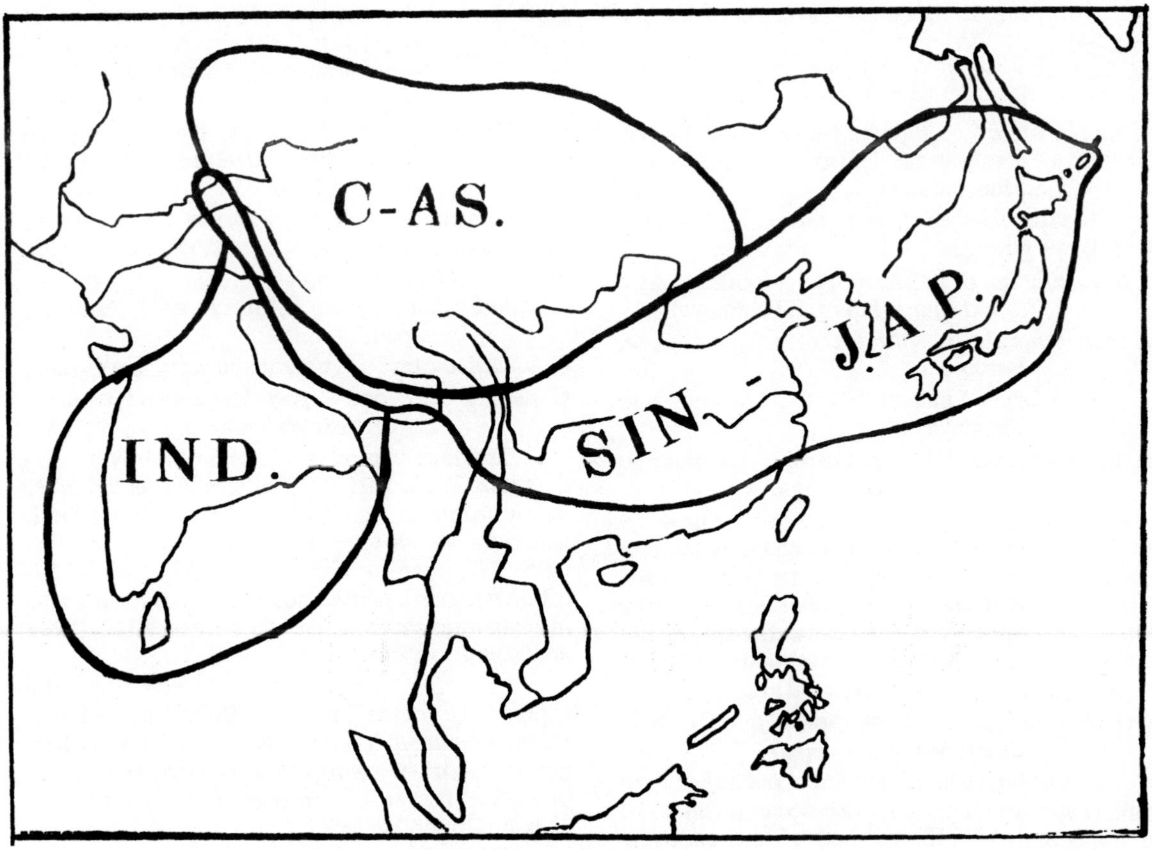

Fig. 53 The floristic regions of India, Central Asia and and Sino-Japanese. (Meusel, Jager & Weinert 1965 IN Dobremez 1972).

into north Asia, E Himalaya and parts of Europe (Fig. 54).

Eupatorium reveesii Wall ex DC native of China also occurs in Japan.

Glaux maritima L. ranges in Europe, N America, Caucasica, Cantral Asia, Japan, Manchuria, etc.

The four species of *Herminium*(Fig. 54) show the following range:

H. angustifolium (Lindl) Bth ex Cl. . southeast Asia (incld China, Japan, E Himalaya)

H. monorchis (1.)Br. widespread in Central Asia, Europe.

H. pugioniforma Lindl ex Hk. f. . . extends into Eimalaya.

H. macrophyllum (D Don)Dandy . endemic to Kashmir, W Himalaya.

Hieracium has six species in Kashmir; of these *H. umbellatum* L. is a central Asian element reaching Japan but *H.virosum* Pall and *H. vulgatum* Fries (all the three are widespread in Europe) do not reach E Himalaya and the heart of Sino-Japanese region (Fig. 55).

Lychnis brachypetala Hort ranges into N Asia, China, Mongolia, Japan.

Potentilla is a large element of our alpine flora with 47 species.Of these 13 also occur in China, 10 in Mongolia,13 in Europe and 13 are endemic. *P.anserina* L., *P. fruticosa* L. and *P. multifida* L reach Japan, also occurring in China and Europe. In *Rumex*, *R. nepalensis* L ranges into Japan, the rest 12 species of Kashmir lean NW-wards, but in *Polygonum* half-a- dozen species range into Japan.

Thymus serpyllum L ssp *quinquecostatus* (Celak)Kita ranges into China, Japan and Manchuria.

(b) SPECIES THAT REACH CHINA BUT DO NOT OCCUR IN JAPAN.

Agrostis canina L. . Europe, Tirich-Mir, N Mongolia, Siberia.

Cimicifuga foetida L. . central Europe-Siberia-Mongolia.

Comarum salessovianum (Steph.)Aschera et Gaertn, . Altai, Alatau, Tien Shan,Pamir-Alai, Tirich Mir, E Tibet, NW China.

Crepis flexuosa (DC)Bth ex Hk. .f.* . . Pamir, Tien Shan, Tirich Mir, outer Mongolia (Lake Baikash- Altai mts), NW China.

Gentiana(Fig. 29,63) shows rich development in Kashmir and has 35 species , of these 14 are endemic, three reach Japan:

G. algida Pall . . Europe, boreal Asia, America, Japan.

G. aquatica Stev+ . . . boreal and arctic north Asia.

G. humilis Stev . and *G. pseudoaquatica* Kuxnez . . boreal Asia and America.

The Eurasian pairs, *G. amarella* in west and *G. acuta* in the east are " interesting as the latter species extends eastwards across the Bering Strait and right across temperate North America" (FBI 5,98). *G. prostrata* Haenke is "fairly widely distributed in north temperate zone and occurring also in temperate south America and Australia" (Dhar and Kachroo 1983).

The following species are common with China: *G. argentea* Royle ex D Don, *G. azurea* Bge, *G. leucomelaena* Maxim (also in Uzbekistan, Kirgiz & Pamir), *G. marginata* (D Don) Grisob, *G. prostrata* Haenke. *G. stracheya* (Cl.)Kita, *G. tianshanica* Rupr (also in Kirgiz).

Hedysarum a north temperate genus is represented by 23 and 19 species respectively in Kirgiz and Uzbekistan but with just six species it has a low profile in Kashmir. *H. cachemiricum* is endemic. The corolla appears to be the most distinguishing feature delimiting some Himalayan species as shown below:

bright red . . *H. sikkimenense*
yellowish . . *H. astragaloides*
red . . . . . *H. cachemiricum*
not known . . *H. kumaonesis, H. falconeri.*

None of the central Asian elements are common to Kashmir:

---

* "*Crepis* is of central Asian origin . . primitive C. *kashmirica* probably moved into western Himalayan region from Central Asia during late Tertiary, It may be a relict species "(cf, Dhar and Kachroo 1983)

+ *G. aquatica* L var *karelinii* Cl is endemic to Ladakh. Is it the same as *G. karelinii* Griseb of Uzbekistan, Kirgiz and Pamir

Hymenolobus procumbens (L.)Nutt ex Torr extends into Europe, Mediterranean, Mongolia. *Holosteum umbellatum* L ranges in NE Europe to N Africa.

Jurinea is yet another genus richly represented in west and central Asia, but poorly so in the Himalaya.

J.ceratocarpa (Decne)Bth . . Kashmir (endemic).
ceratocarpa var *depressa* Cl ex Hk f. . W Tibet-Kashmir (endemic).
macrocephala G Boiss . . Kashmir-Kumaon.
modesta Boiss. . Pakistan (Peshawar), Afghanistan.

In Uzbekistan 33 and in Kirgiz six species occur. The genus appears to be absent from Pamir proper. In the USSR 163 species!are recorded, (of these 70 are described by Iljin!).

Meconopsis is essentially a genus of the Himalaya which extends into Europe and is discontinuous. In India the species are classed in two groups as follows:

1 subalpine . with scapes radical and 1-fid, most of these are rare, excusively in Sikkim (& Nepal) and one in Kashmir- Paksistan.
2 subalpine to temperate . . with stems leafy and flowers racemed or panicled, occurring in Kashmir to Sikkim.

Interestingly *M. aculeata* Royle alone belonging to group 2 is prickly and resembles in this respect the prickly *M. horridulum* HK. f. & T. of group 1. Both have capsules also prickly and flowers blue purple. Perahps this lead Hooker f & T to state that" possibly a state of *M. aculeata*" (FBI). *M.neglecta* G Taylor is yet another member of group 1 but is restricted to Kashmir-Pakistan.

Taylor regarded *Meconopsis* as consisting of one western and ca 40 species in the Sino-Himalayan mountain system. Some taxonomists include the western north American species also in this system. However, the genus is regarded as a true Sino- Himalayan element. *M. aculeata* alone occurs in Kashmir alpines and hardly 3 extend beyond W Himalayan.

Myricaria squarrosa Desv ranges in Crimea-Caucasus, Iran, Tirich Mir, Pamir, Asia, Tien Shan, Altai and eastwards to Dahuria and N China. Four species of *Pimpinella* occur in Kashmir, except for *P. diversifolia* (Wall.) DC which occurs in Afghanistan, Pamir, E Himalaya, Taiwan and Japan; other species range north-westwards.(Fig. 62).

Sedum heterodontum HK. f. & T. ranges in Tirich Mir, Pamir Alai, Tian Shan and Monoglia. *Sibbaldia cuneata* Kunze into Afghanistan, Pamir, ? Tibet, west and central China. *Torularia humulis*(CAM)OES is an element of Afghanistan-Mongolia-China.

Of the local species of *Youngia*, *Y. japonica* (L.)DC is widespread in southeast Asia (Malaysia, Thailand, Vietnam), Occurring throughout India and ranging southwards into Austraiasia. *Y. tenuifolia* (Willd) Babc ssp *diversifolia* Babc & Steob ranges from Siberia-Mongolia-China to Central Asia. Thus the local taxon provides a outpost from Australasia to Siberia(Fig 29).

Sino-Himalayan elements

Li (1944) recognised a Sino-Himalayan region and the Tibetan-Plateau region; the former includes the eastern and northeastern portions of the Great Tibetan Plateau (ave. ele. 3,000 m) revealing a" close kinship with the Himalayan flora and showing extensive development of alpine genera like *Rhododendron, Primula, Meconopsis, Anemone, Saxifraga, Gentiana, Pedicularis.* He advocated separation of western Chinese elements from the Himalayan elements as many of these are not common". He also did not favour the use of the term "Sino-Himalayan".

The position of the aforesaid genera with respect to our area is as follows:

Rhododendron (3 sps) . 2 reach E Himalaya.
Primula (21 sps).. 4 reach China, 4 E Himalaya.
Meconopsis (1 sps) . . endemic.
Anemone (10 sps) . . 1 reaches E Himalaya, 1 Central Asia.
Saxifraga (25 sps) . some elements reach E Himalaya and Central Asia.
Gentiana (35 sps) . . 2 reach Japan, 8 China.
Pedicularis (27 sps) . . 4 reach E Himalaya, 4 Central Asia.

In the Tibetan Plateau region is inculded most of the political Tibet, western parts of Sinkiang and largest parts of Chinghai and the mountain chains and valleys that criss- cross from west of east and have higher altitudes and latitudes than Sinkiang and Mongolia, "Whereas the Sino-Himalayan region holds a rich proportion of endemic elements with a little mixture along the southern slopes of the Indo-Malayan elements".

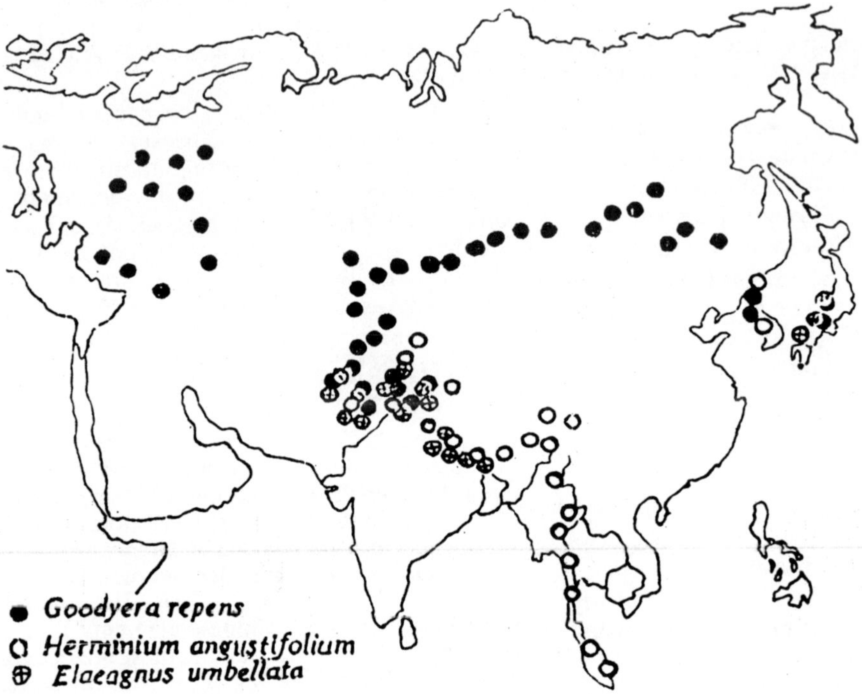

Fig. 54 : *Goodyera repens* is widespread whereas *Herminium* ranges into SE Asia as well as China and E Himalaya, the former is one of the taxa that bounces over the Sino-Himalayan and Japano-Himalayan regions. *Goodyera* abounds in N temperate and tropical region Asia, *Herminium* is mainly temperate Asia. *Elaegnus umbellatus* ranges into E Himalaya and W China, a native of Japan but absent in mainland China; through *orientalis* it extends into Europe.

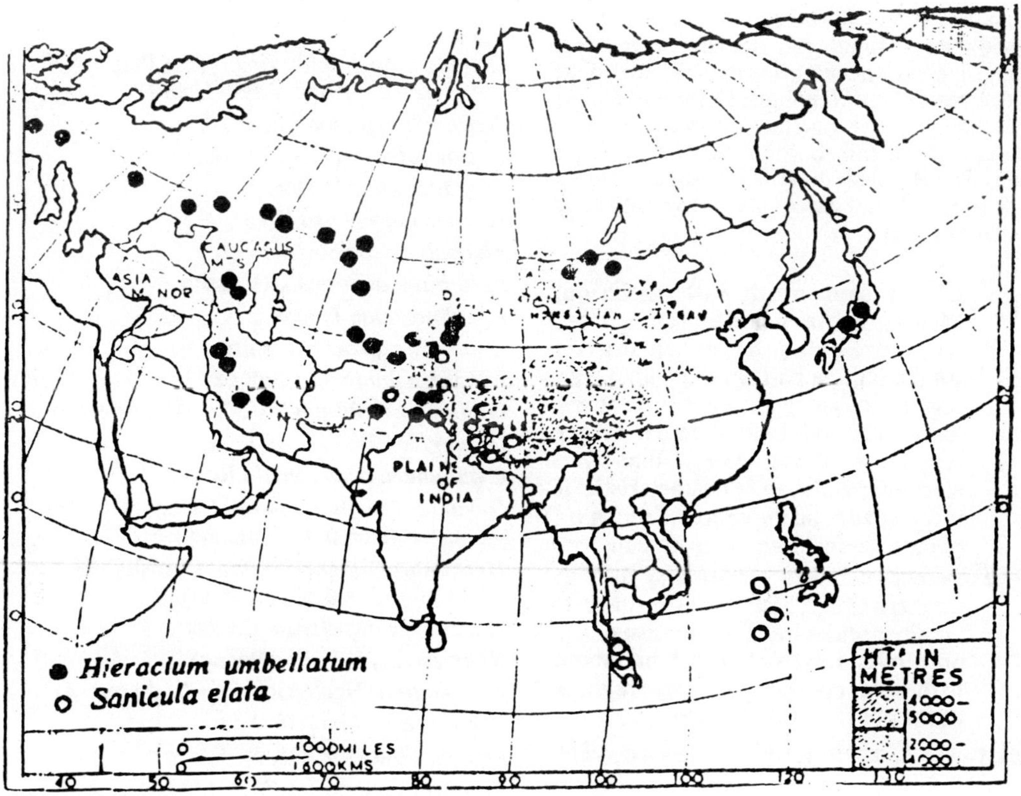

Fig. 55 : *Sanicula elata* is widespread in east and west but in India it is confined to the Himalaya, and next occurs in SE Asia. *Hieracium umbellatum* in also widespread but does not occur in C and E Himalaya and (?) China. Its presence is Japan is interesting!

Japano-Himalayan elements

Kanai (1963) while discussing distribution pattern of Japano- Himalayan elements stated that a number of such species (Table IV) extend in temperate forests of the Himalaya, China and Japan; a few of taxa we find range into Kashmir as well; and in some cases this is their farthest western extension. He divided these Japano-Himalayan plants into two groups, " especially concerning their northern ranges (Fig. 56).

Group 1: absent from north eastern China, Manchuria, Amur and especially from Korean Peninsula, excepting its southern most part. In the Himalayan region they are represented in a narrow E-W belt and hold " much larger proportion of the Japano-Himalayan elements" and may be characteristic. It is regarded as an important subdivision of the eastern Asian Temperate region.

Group 2: present in Korea proper and also in northern China, Manchuria or Amur. It is rather widespread, some taxa occur both in Asia and N. America, and some in the temperate and subalpine regions of the northern hemisphere.

Kashmir-W China-E Himalayan elements

Plants common to these areas are enumerated below.

Acer cappadocicum Gled

Anaphalis nepalensis (Spr.) Hand.-Mzt.

A. timmua D Don (+ Vietnam, Korea, America, Kamachatica).

Anemone rivularis Buch.-Ham. (+ Karakoram).

Arabidopsis mollissima (CAM) Busch.

Artemisia roxburghiana Wall ex Boiss.

A. dubia Wall.(+ Japan).

Barbarea intermedia Boreau ?

Braya oxycarpa Hk. f. &. t.

B. tibetica Hk f. &. T.

Bupleurum candollei Wall ex DC

Campanula colorata Wall ex Roxb

Cardamine macrophylla Willd (+ Monoglia, NE USSR).

Circaea pricei Hayata (circumpolar, Afghanistan).

Circium wallichii DC

Clematis buchaniana DC

C. connata DC

Cynanchium vincitosxicum (L.) Pers (+ N Africa, Europe, Asian USSR).

Desmodium podocarpum DC

D. tilaefolium (D Don) R Br.

Dipsacus mitis D Don

Deutzia corymbosa R Br

Elscholtzia strobilifera Bth

Epilobium amurensis Hausskn.

E. cylindricum D Don

Euphorbia stracheyi Boiss.

Fragaria nubicola Lindl ex Laccita (+ USSR).

Galium aperifolium Wall. (+ Afghanistan)

G. elegans Wall *(+ Afghanistan)*

Cynanthera trichophylla Royle

Geranium wallichianum D Don et Sweet (+ Afghanistan, Korea, C Himalaya).

Herminium angustifolium (Lindl) Bth ex CL (+southeast Asia).

Hypericum nepalensis Choisy

Kobresia capillifolia (Decne) Cl

K. royleana (Nees) Boeck (+Afghanistan)

Lactuca dolichophylla Kita.

Lancea tibetica Hk. f. &. T.

Lonicera asperifolia (Decene) Hk. f. &. T.

L. quinqecularis Hardw.

Myricaria rosea WW Smith

Papaver nudicaule L. (+Siberia, Mongolia, Europe).

P. dubium L. (+ temperate, alpine, wide).

Pegaeophyton scapiflorum (Hk. f. &T) Marq et Shaw.

Pilea umbrosa Wedd.

Polgonum filicaule Wall ex Meisn.

P. molle D Don (+ Vietnam)

Ribes glaciale Wall

R. emodense Rehder

R. alpestre Decne ex Jacq

Saxifraga pallida Wall ex DC

Senecio chrysanthemoides DC (+ Afghanistan)

Sibbaldia cuneata Kunze

S. microphylla (D Don) Hand.- Mzt.

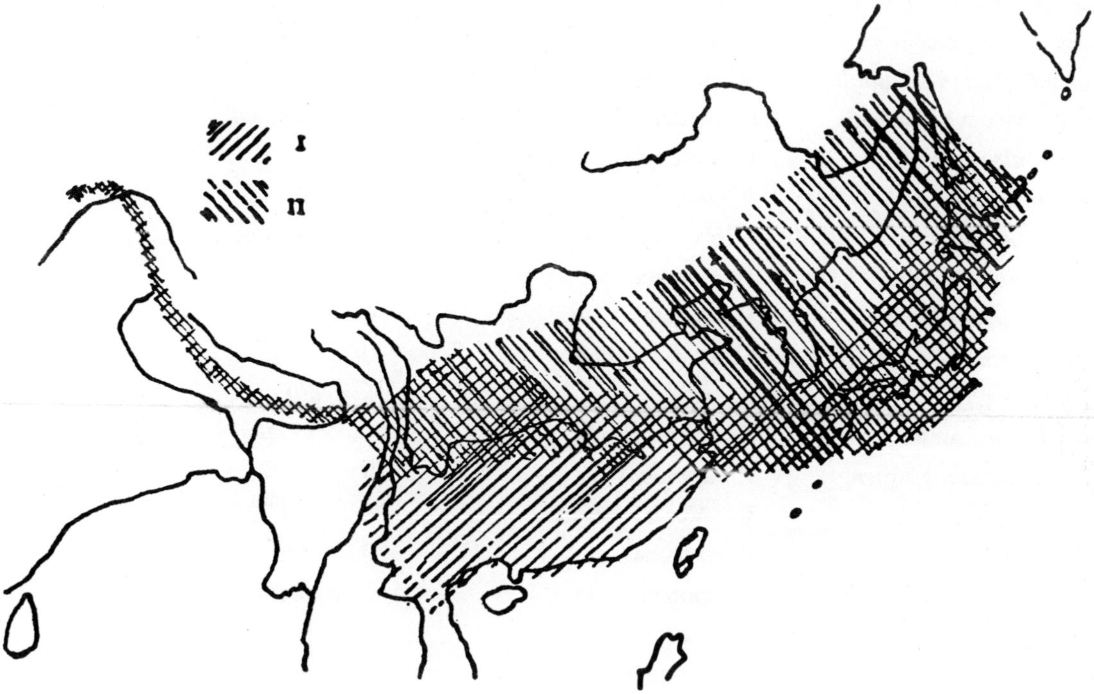

fig. 56 : The range of Japano-Himalayan elements. (cf Kanai 1963).

Sophora moorcroftiana (Bth) Bth ex Baker.

Stellaria alsine Grimm (+Europe, central America,Mediterranean,Scandinavia, Asia Minor, N Asia).

Thylacospermum caespitosum (Camb) Schis. (+ boreal Asia).

Torularia humilis (CAM)OES(+ boreal Asia, America).

Valeriana jatamansi Jones (+ Afghanistan)

Viola biflora L.

Winkstromia canascens Meissn (+ Afghanistan).

Table IV Japano-Himalayan* elements in relation to Kashmir.

Himalayan plants	corresponding* Japanese plants	corresponding** Kashmir plants
1. Adenocaulon himalaicum	A. himalaicum	0
2. Aucuba himalaica	A. japonica	0
3. Boenninghausenis albiflora	B. a var *japonica*	*B. albiflora*
4. Cardiocrinum giganteum	*C. cordatum*	*C. giganteum*[1]
5. Clinotonia alpina	C. a var *udensis*	0
6. Enkianthus deflexa	E. *campanulatus*	0
7. Geranium nepalensis	G. n var *thubergi*	*G. nepalensis*
8. Helwingia himalaica	H. *japonica*	0
9. Houttuyria cordata	H. cordata	0
10.Hydrangea anomala	H. a ssp *petiolaris*	0
11.Lyonia ovalifolia	L. *neziki*	*L. ovalifolia*[2]
12. Oxalis acetosella	*O. acetosella*	*O. actosella*
13. Panax pseudo-gingseng	P. *Japonicus*	0
14. Peracarpa carnosa	P. c Var *circaeoide*	0
15. Picris[3] *formosa*	P. japonica	0
16. Skimmia wallichii	S. japonica	*S. laureola*
17. Stachyurus himalaicus	S. praecox	0
18. Swertia himaculata	S. himaculata	*Shdiverse species*
19. Tiarella polyphylla	T. polyphylla	0
20. Mimulus nepalen sis	M. n var *japonica*	*M. strictus*[4]
21. Magnolia globosa	M. sieboldii	0

* Kanai (1963), ** Dhar & Kachroo 1983 .

1 Karna, Muzzafarabad; 2 Poonch, Udhampur; 3 *Picris* pp is now *Lyonia*; 4 Poonch, Muzzafarabad .

FLORISTIC AFFINITIES

Abbreviations used

Afg, Afghanistan
Afr, Africa
Amr, America
Ch, China
Eur, Europe
H, Himalaya (NW, northwestern, E, eastern, C central, W, western)
Hind, Hindukush
Jap, Japan
Kar, Karakoram
Kash, Kashmir
Manch, Manchuria
Mong, Mongolia,
N, north
NZland, New Zealand
Pak, Pakistan
P-A, Pamir Alai
Sib, Siberia
S, south
SE, southeast
Tb, Tibet
Tur, Turkistan Turkmenia
CA, Central Asia
W, west
WS, widespread

Kashmir as a part of NW Himalaya, and a part of Hindukush has dry climate and may belong to the Ancient Mediterranean Region (AMR). But the Mediterranean region of today has considerable monsoon influence and Raven (1971) has defined it as " a mediterranean climate is one in which the summers are dry and hot and the winters are cool and moist. . Unlike most of the temperate regions, plant growth is active during winter, inactive and often completely suspended during summer" (Fig.57). Both the above types are different from the desert climate of the mountains of Afghanistan. Northwards both Kashmir-Hindukush and Afghan regions just abut on the Tibetan region.

It is agreed that the alpine and subalpine Himalayan province belongs to Central Asia (Engler 1899). Dhar (1978) also drew attention to this affinity with respect to Kashmir Himalaya. Engler (1879) discussed the southwesterly trend of flora of Central Asia towards Asia and Europe by drawing inferences from flora of high mountains of the Himalaya, Pyrenees, Alps, Altai, etc. He also suggested a trend of migration from Siberia to Caucasus and Alps through a southward-south-west direction to the Mediterranean mountains. In this connection the reader's attention is drawn to the principal mountain chains in Asia and the trends of Tertiarty plant migration (Fig. 58). This trend is exemplified by a few taxa ranging into Kashmir, bouncing wholly or partly, as it were, over the E Himalaya-Sino-Japanese triangle.

Taxa ranging into Kashmir

(a) BOUNCING WHOLLY

Artemisia salsoloides Willd var *paniculata* Hk. f, *Ceratium cerastioides* (L.) Britt (+ Iran, N Amr), *Lloydia serotina* Reichb (+Turk,NAmer!), *Kobresia macrantha* Boeck (+ Iran), *K. schoenoides* Boeck (+ Iran), *Medicago falcata* L., *Myosotis* cf *asiatica* Schischk et Serg, *Potentilla bifurca* L.(+ Iran, Turk).

(b) NOT REACHING JAPAN, BUT COMMON TO SIBERIA AND WCHINA (and WH or H)

Arabidopsis mollissima (CAM)OES var *dentata* OES(+ Iran, H), *Artemisia turczaninowiana* Bess (+ WH), *Aster flaccidus* Bge ssp *flaccidus* Grierson (+WH), *Astragalus frigidus* (L.) A Gray,

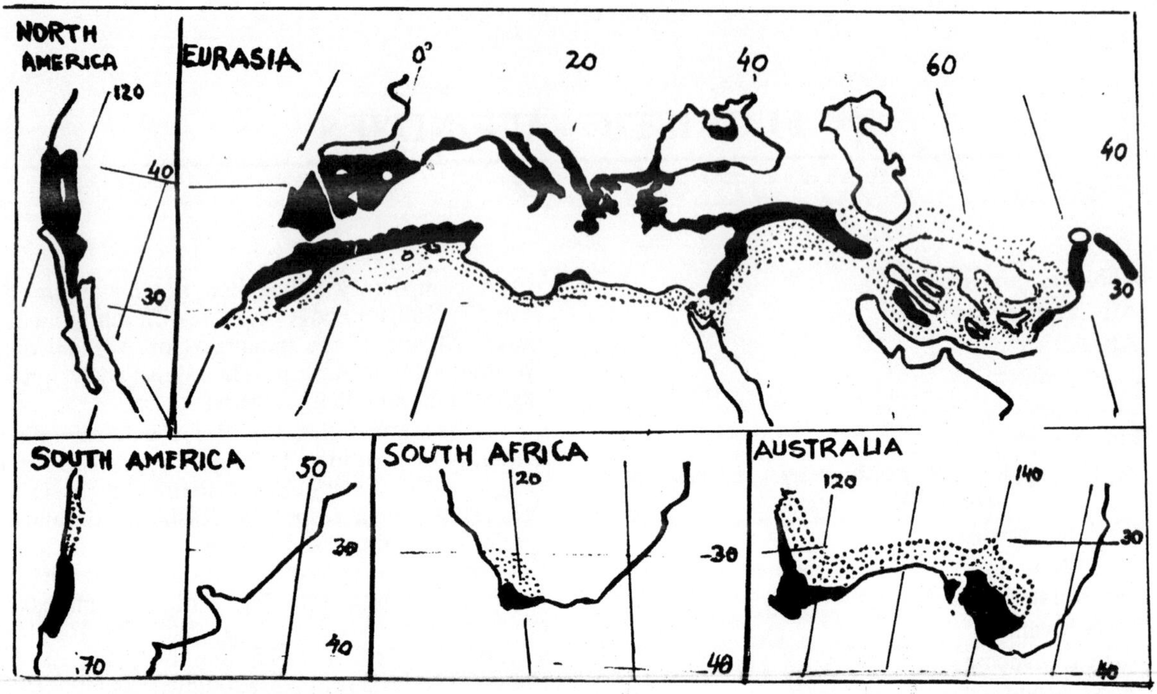

Fig. 57 The five regions of the world with a Mediterranean type of climate (cf Raven 1971).

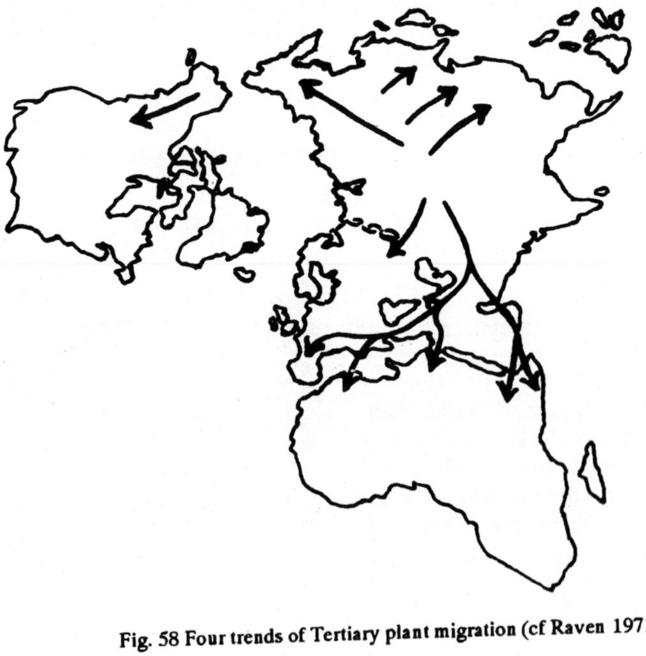

Fig. 58 Four trends of Tertiary plant migration (cf Raven 1971)

(+NWH, NAmer!), *Carex pseudofoetida* Kuk (+Iran, NWH), *Chorispora sibirica** DC, *Elymus dasystachys* Trin (+ NWH), *Geranium collinum* Steph ex Willd var *glandulosum* Ldb (+ Iran, WH),*Hippophae rhamnoides* L (+Iran, WH), *Melilotus officinalis* Lam (+ WH), *Parnassia ovata* Ldb (+H), *Poa bulbosa* L. (+ NWH, A Afr!) *Polygonum viviparum* L (+H), *Saxifraga flagellaris* Willd ssp *euflagellaris* Engl et Irn, (+ H), *S. sibirica* L var *eusibirica* Engl et Irn (+ H), *Silene tenuis* Willd† (+ WH), *Stellaria graminea* L (+ WH).

(c) NOT EXTENDING TO CHINA BUT COMMON TO JAPAN

Carum carvi L (+ Iran, H, NAfr, N&S Amer, NZland), *Papaver nudicaule* L ranges from W Europe-N China-Mongolia-Siberia, but *P. dubium* does not occur in E Himalaya.

Papaveraceae it may be recalled is an element of S Africa and Australasia. *Swertia* (13/12/75%)** may also belong here as one of its species (*S. marginata* Shrenk) reaches Siberia and of the rest only a few reach western Himalaya and none Sino-Japanese region.

Salsola (3 species) has wide distribution as shown below:

S. kali L. . throughout Europe.

S. collina Pall . . Iran and Siberia.

S. ruthenica Iljin . . Himalaya

Prunella vulgaris L of Siberia occurs in Kashmir-Sikkim, Europe, N Africa, Asia Minor, Afghanistan, Iran, Caucasus and Central Asia. (Is it a Irano- Turanian or Mediterranean element ?).

Affinities to adjoining regi

We find that in quantitative terms, the number of species that occur in Kashmir Himalaya are practically the same as those in the adjoining NW and northern regions of Asia (Table V ; Fig. 59)

The affinities to Iran are not considerable and with E Himalaya not better; whereas relations with Siberia appear of consequence in view of the "isolation of Angaraland in E Siberia (Sahni 1935) at a time when it was separated from the Eur-american coal flora in the North Atlantic Continent by a broad sea later to be closed as China. Angaraland and Europe drifted together in the Cretaceous to form Eurasian Continent. Between the northern continents and Gondwanaland was a circumglobal sea, the Tethys (Melville 1981).

The distribution of alpine and nival species of Hindukush (Fig. 60) brings forth intimate relationship between this region and Kashmir on one hand and Central Asia on the other; and Breckle (1974) has stated that those plants which occur at highest elevations have widest distribution because of the micro- climatic factor which ensures "intensive seed propagation and propagation by vegetative parts". Thus the maximum affinity of Kashmir is with the eastern Hindukush understandably least with central Afghanistan. Hindukush shares 78 % of its flora with 'Little Pamir'and 55 % with Kashmir (Fig. 61).

Families and genera with maximum species

The families of flowering plants with maximum species in Kashmir and their relative position in India as a whole is given below: (arabic numerals within brackets indicate rank in Indian flora):

Family	*Kashmir*	*India*
Compositae	334	697(1)
Leguminosae(s lato)	172	862(2)
Cyperaceae	125	?
Rosaceae	104	255
Cruciferae	97	174
Ranunculaceae	92	162
Scrophulariaceae	85	274
Labiatae	81	421(6)
Rubiaceae	21	555(3)
Acanthaceae	4	508(4)
Euphorbiaceae	14	442(5)

* not occurring in the Himalaya.

† However, *Silene vulgaris* (Moench) Garcke reaches Japan, Eurasia, N Africa, and N & S America

** total species in Kashmir/alpine-subalpine species/endemism percentage. (explanation for such figures where-ever these may occur in this text).

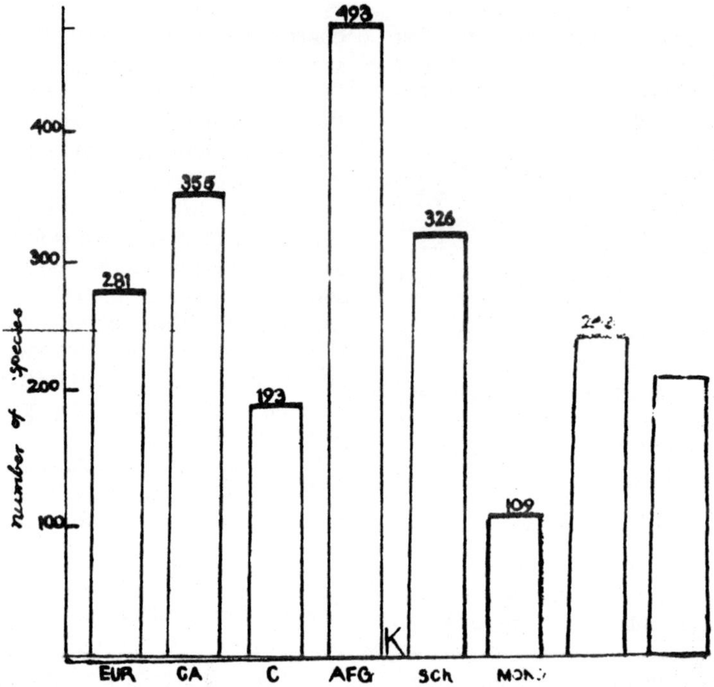

Fig. 59 : Affinities of Kashmir flora with some adjoinig regions. *EUR*, Europe; CA, Central Asia; C, Caucasus; *Afg*, Afghanistan; K, Kashmir; *Sch* S China, *Mong*, Monglia; *Sib*, Siberia; *EH*, East Himalaya.

The families occupying 3rd, 4th and 5th rank in India as a whole are poorly represented in Kashmir and their distributional pattern so far as local elements are concerned is striking in being largely restricted:

Rubicaceae (555/21), a largely tropical family, only *Galium* is widespread.

Acanthaceae (508/14), *Strobilanthes* alone reaches E Himalaya.

Euphorbiaceae (442/14), only one species is widespread!

Another striking aspect is that most of the larger families in India are tropical or belong to moist temperate regimes; thus their representation at the genus level in our area is also poor (Table VI). The families Euphorbiaceae, Acanthaceae, Asclepiadaceae, Capparidaceae and Celastraceae each having just a single genus in our area!

We may now consider the genera with maximum species in our area. They are largely restricted to Central Asia-SW China axis (Table VII) and show feeble extensions eastwards!

Genera with single species

The genera with single species in our area number ca 201 with 36 species common to Central Asia where 378 such species occur in Uzbekistan and 302 in Kirgiz. Such species common are:

Uzbekistan and Kirgiz. . . .149
Kashmir-Uzbekistan 15
- Kirgiz 29

With respect to their distribution pattern these show affinities with Irano-Turanian as well as Sino-Himalayan provinces as shown below:

Mediterranean . . *Cimicifuga foetida, Hymenolobus procumbens, Capparis spinosa, Agrimonia eupatoria.*

Iran . . *Tauscheria lasiocarpa, Petrohagia alpina, Lapyroidictilis holosteides, Bieberste inia odora, Microsisymbrium minuti florum* (+ Afghamistan).

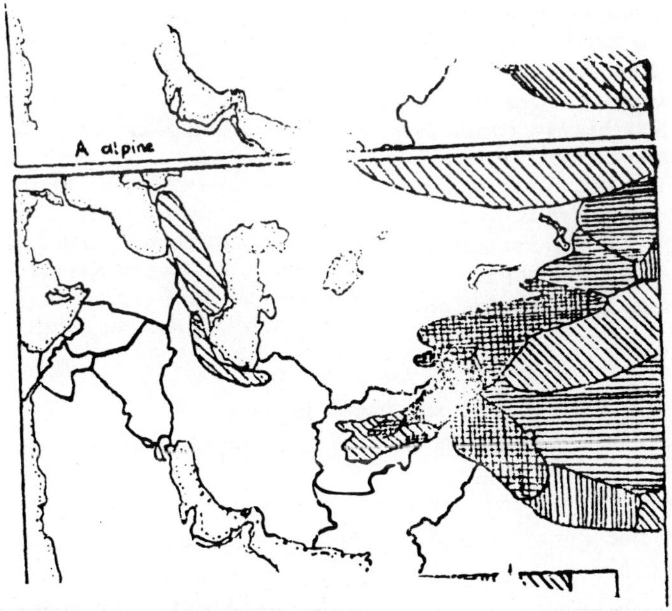

Fig. 60 Affinities of flora of Hindukhush with adjacent regions: plants growing above 4000 m(A), above 5000 m (B), in both cases the number of plant species was taken as 10 %, alpine 377; nival 36. (cf Breckle 1974).

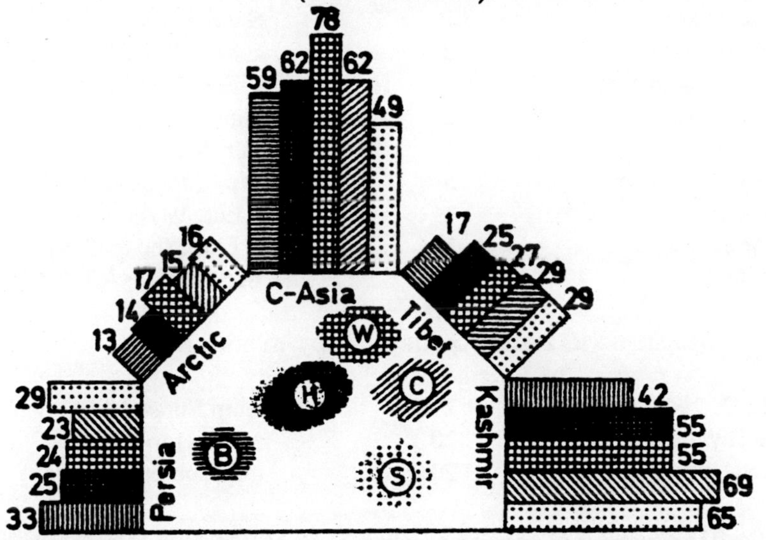

Fig. 61 The relationship of the flora of mountain parts of the Hindukush to the flora of adjacent mountain regions (cf Breckle 1974). (species found at 4000 m and above only considerded. Mountains considered:/HINDUKUSH: B, Koh-e-Baba in C Afghanistan; H, W,N,C Afghanistan Hindukush; W, Wakhan range (Little Pamir), C, Chitral and E Hindukush; S, Safed Koh. PERSIA: Iranian highlands, Turkey, E Madeterranian mts. ARCTIC: Arctic and boreal elements, Siberian elements. C ASIA: central asiatic mts, Pamir, Alai, Tien Shan, Tarbagatai. TIBET: Inner asiatic

East Himalaya . . *Archyospermum primulifolium, Pagaeophyton scapiflorum, Torularia humilis, Gerbera nepalensis, Aralia cachemirica, Agrimonia eupatoria.*

China . . *Hymenolobus procumbens, Pagaeophyton scapiflorum* (S China), *Torularia humilis, Sophora moorcroftiana, Filipendula vestita, Eupatorium reevesii, Gerbera nepalensis.*

The genera with single species common to same areas in Central Asia are the following.

Some of the other species common to Iran are the following:

Arabidopsis mollissima (CAM) OES (WSib, CA,H,), *Ceratoides papposa* (CA, Sib), *Couringia persica* Boiss (Caucasus, Armenia, Afg), *Hippophae rhamnoides* L (WCh, Eur, Sib), *Lappula barbata* Griecke var *cariensis* (Boiss) Brand (WH), *Lepyrodiclis holosteoides* (CAM) Fenz (Turk, WCh), *Macrotomia perennis* Boiss (WH,WCh), *Mathiola flavida* Boiss (Hind, Kar, NWH), *Myricaria squamosa* Desv (Caucasus, CA, NCh), *Pedicularis pycnantha* Boiss (Afg, Turk, CA), *Perovskia abrotanoides* Kar (W Tb, *not H*), *Prunus jacquemontii* Hk. f. (Afg, P-A, WH), *Scrophularia variegata* MB (Caucasus, Afg).

Species/Family	number species in the genus	distribution worldwide
SPECIES COMMON TO KAZAKHSTAN, UZEKISTAN, KIRGIZ, TADJIKSTAN AND KASHMIR		
Apium graveolens , Umbelliferae	1	Eur to As, N & S Afr
Capsella bursa-pastoris, Cruciferae	5	temp subtrop, cosmop weed
Conium maculatum, Umbelliferae	4	N temp, Euras, S Afr
Cydonia oblonga, Rosaceae	1	E AS Min, N Persia, C As
Datisca cannabina, Datiscaceae	1	Med-Himal & C As, SW US NW Mex
Dactylis glomerata, Gramineae	5	temp Euras
Diarthron vesiculosum, Thymelacaceae	2	temp As
Drabopsis verna, Cruciferae	1	Sw & C As
Euclidium syriacum, Cruciferae	2	E Med-C As & NW India
Gastrocotyle hispida, Boraginaceae	2	E Eur-C As
Hippuris vulgaris, Hippuridaceae	2-3	almost cosmop
Hymenolobus procumbens, Cruciferae	5	Eur ,Med C As, Austr, NAM, Chile
Peganum harmala, Zygophyllaceae	5-6	Med-Mongolia, S US, Mex
Sclerochloa dura, Gramineae	1	S Eur, W As
Turgenia latifolia, Umbelliferae	1	C Eur, Med to C As
Turritis glabra, Cruciferae	3	Eur-Jap mts, Afr, ? N Am
SPECIES COMMON TO KASHMIR, KAZIKHSTAN, KIRGIZ AND TADJIKSTAN		
Coelog lossum viride, Orchidaeae	2	temp Euras, N A
Thymaela passerina, Thymelaeaceae	20	Med, temp As
Tribulus terrestris, Zygophyllaceae	20	trop, subtrop (cal-trop)
SPECIES COMMON TO KASHMIR. KAZAKHSTAN, UZBEKISTAN AND TADJIKSTAN		
Buchingera axillaris, Cruciferae	1	Armenia & Iran to C As
Tauscheria lasiocarpa, Cruciferae	2	C As

Table V. Affinities of Kashmir flora with that of the adjoining regions,based on number of alpine and subalpine species common (cf Dhar and Kachroo 1983b).

	Dicots		Monocots	
Area	species	%ge	species	%ge
Afghanistan	440	31.31	53	25.58
Central Asia	312	22.20	43	21.00
Southwest China	289	20.56	37	18.40
Europe	250	17.79	31	15.00
Siberia	216	15.37	32	15.60
Caucasus	174	12.38	19	09.26
East Himalaya	173	12.10	45	02.95
Iran	150	10..67	08	03.90
Mongolia	106	07.54	03	01.64

Of Polygonums only *P. paronychoides* occurs in Iran. Some other species common to Karakoram show the following range:

P. affine Don . . EAfg- WTb,

paronychoides CAM . . Iran, Caucasus, WH, P-A, Syr Darya.

rupestre K et K . . WCh, Tb, Altai, WH.

viviparum L . . wide, temperate of both hemispheres.

Endemic diversity

An interesting aspect of endemism is that it is associated with alpinism and in certain cases all the endemics are alpine (Tables VII, VIII).

A glimpse at Tables VII-X reveals quite a diversity as regards endemism in flora of Kashmir and some adjoining floral provinces. Considering Afghanistan and Kashmir alone, families heavily studded with endemism are nearly the same. However our families with 49 % or above endemism are rather distinctive. Campanulaceae, Umbelliferae, Scrophulariaceae and Cruciferae which are mildly endemic have more than 50 % endemism in Kashmir flora. In contrast to Iran we find that Chenopodiaceae is rather a poor element (19.42%) and Boraginaceae and Labiatae have less than 40% endemism (Table IX) in our area.

Of the first major endemic families of India (endemism 80% and above) as a whole Balsaminaceae has 43% and Fumariaceae 52% endemism and the rest fare very poorly in Kashmir: Papaveraceae ca 17%, Ericaceae 20% . This is nearly true of Labiatae (27%).

Evidently the richness in endemism of our major taxa and those of Afghanistan (Table X) only indicate that speciation has occurred in the 'recent' past. Same is true of Central Asia.

However one may not agree with Chatterjee (1939) who traced Indian crucifers from the Mediterranean, because of the fact that great diversification has occurred in the family in Central Asia where crucifers found enough scope for speciation on the high mountains (see also Dhar 1978). Our crucifers have 92 % alpinism. 44 % are related to Central Asia, just 13 % to eastern Himalaya and 27% are endemics. (In India as a whole Chatterjee recorded 56 % endemism in the family).

Impatiens is yet another example to show that speciation could have occurred (in this genus) even in Kashmir, two of our species (*I. gigantea* and *I. thompsonii*) have western limits as well.

Table VI. Some major families of Indian flowering plants: relation of genera in India to those of Kashmir.

Family	India : Kashmir	Family	India : Kashmir
Compositae	126 : 50	Rutaceae	24 : 2
Leguminosae	112 : 17	Urticaceae	20 : 3
Euphorbiaceae	70 : 01	Saxifragaceae	17 : 3
Labiatae	69 : 16	Gentianaceae	15 : 5
Rubiaceae	68 : 03	Campanulaceae	13 : 3
Scrophulariaceae	54 : 11	Capparidaceae	10 : 1
Acanthaceae	50 : 01	Celastraceae	10 : 1
Asclepiadaceae	49 : 01	Thymeliaceae	10 : 2
Geraniaceae	34 : 01		

Table VII. Range of genera* with maximum species in Kashmir Himalaya based on Dhar & Kachroo 1983a

Genus	Total species	Alpine-subalpine species	Endemism(%)	Remarks
Taraxacum	59	50	50	
Astragalus	47	47	53	Eur, Ch
Potentilla	47	36	26	widespread (ws)
Gentiana	32	31	51	4ws,10 N Asia, 4CA, 4Ch
Artemisia	31	29	23	6 Afg
Nepeta	28	25	59	4 CA, 7 Afg
Pedicularis	27	25	53	ws, NAmer
Saxifraga	25	25	?	1 WH
Saussurea	31	24	80	ws
Corydalis †	25	24	55	none common with Eur, 4 EH, 4 Ch, Caucasus
Ranunculus	27	23	00	ws
Polygonum	41	20	51	13 ws, 5 EH
Veronica	19	51	26	ws, 2 EH
Lonicera	18	18	?	3 EH, 4 Ch
Primula	21	13	61	3-4 EH, Ch
Draba	18	17	47	
Epilobium	21	16	25	ws
Lactuca	16	15	27	2 EH
Sedum	19	12	33	
Geranium	18	12	23	1 CH
Bupleurum	17	12	48	2 Ch, 2 EH
Berberis	16	09	66	mostly neoendemics, none beyond Afg.

* All genera mostly lean towards north or NW-wards, except in addition to where shown. Our species in the following genera do not reach EH: *Aconitum, Arabis, Gentiana, Nepeta, Oxtropis, Skimmia.*

† The main development of *Corydalis* has taken place in Central Asia and the Himalaya from

Table VIII. Relationship between alpines and endemism in predominantly alpine families of Kashmir Himalaya (cf Dhar & Kachroo 1983 b).

Family	Alpine sps % of total flora	Endemism % of species + vars only	total flora species
Primulaceae	58.13	39.53	30.23
Saxifragaceae	57.14	53.57	32.14
Fumariaceae	56.00	52.00	52.00
Dipsacaceae	50.00	50.00	50.00
Scrophulariaceae	47.05	44.72	37.65
Compositae	37.72	39.82	33.82
Parnassaceae	33.33	33.33	33.33
Gentianaceae	32.72	49.09	32.73

An interesting feature of genera is that in Afghanistan *Dionysis, Acanthophyllum, Cousinia* and *Acantholimon* are still in the process of speciation whereas in Kashmir the first such major taxa are *Taraxacum, Saussurea* and *Primula. Erysimum* which has 36% and Plumbaginaceae 75% endemism in Afghanistan have reached a dead-end in endemism in Kashmir (0%) but *Campanula* (15%) is rich in endemics (47%) in our area.

Sino-Japanese-Himalayan elements

Among the affinities with the Sino-Japanese-Himalayan elements we may refer to such interesting genera as *Gentiana, Geranium, Nepeta, Potentilla* (Fig. 62) *Salsola collina* Pall and *Herminium monorchis* R Br also occur in Kashmir Himalaya and Japan; and *Thalictrum foetidum* L in Mediterranean, WChina and Japan.

Our plants common to China, Japan and Siberia are: *Iris ensata* Thunb (WH), *Plantago depressa* Willd var *eudepressa* Ganjeschin (WH), *Potentilla multifida* L. (H,NAmer !), *Salsola collina* (H), *Thalictrum foetidum* (Iran,H).

The discontinuous Hydrangaceae (distr. America and Eurasia) and Parnassaceae a cool north temperate element without extensions into south show the following pattern:

Hydrangaceae
Deutzia . . *corymbosa*, EH, WCh. *staminea*, EH
Philadelphus tomentosa, EH
Parnassiaceae
Parnassia palustris, WS, CAmer, Ch, Jap.
cabulica, Afg. *nubicola* ssp *occidentalis*, EH, Afg
ovata, CA, Sib, WCh, Mong.

Nepeta and *Geranium* referred to above though mainly with NW and W-extensions have a species in each ranging into WChina; and the absence of our elements of *Nepeta* in EHimalaya is significant !

Cuscuta mainly temperate is widespread with gaps and shows diverse extentions as shown below:

3 species. . WS, Eur,N Asia, Iran, Afg.
1 *(reflexa)* sps . . WS, Indian subcontinent, Ceylon, Malaysia, Java, WCh, CA.
2 sps . . . . restricted, CA, Afg.

Galium (10 sps) is one element which has no affinity with Central Asia, but perhaps is a true element of the Sino-Japanese province.

2 sps *(boreale, cordifolia)* . . Korea, Jap. Vietnam.
1 sps *(cordifolia)* . trop Afr, India.
7 sps . . . . China, EH.

Let us consider *Corydalis, Carex, Primula* and *Potentilla* (Figs. 63) *Carex* links Australasia, N America and Africa; yet major elements of our flora

Table IX. Diversity in endemism of flowering plant families of Kashmir and some adjoining provinces.

Iran[1]	Afghanistan (%)	Kashmir (%)	
Chenopodiaceae	Plumbaginaceae 73		
Cruciferae	Compositae 40	Adoxaceae 100	Balsaminaceae 91
Leguminosae	Leguminosae 40	Balsaminaceae 60	Ericaceae 90
Umbelliferae	Labiatae 35	Berberidaceae 64	Fumariaceae 73
Boraginaceae	Primulaceae 35	Cruciferae 55	Saxifragaceae 96
Labiatae	Campanulaceae 33	Fumariaceae 55	Rosaceae 70
	Umbelliferae 30	Umbelliferae 54	Rubiaceae 67
	Boraginaceae 30	Gentianaceae 54	Labiatae 62
	Caryophyllaceae 30	Violaceae 54	Papilionaceae 57
	Scrophulariaceae 30	Saxifragaceae 53	Cruciferae 52
		Scrophulariaceae 52	
		Compositae 52	
		Campanulaceae 52	
		Celastraceae 50	
		Aceraceae 50	
		Valerianaceae 50	
		Dipsacaceae 50	
		Leguminosae 49	

1. dominant families %ge not known.

2, Chatterjee (1939) included in India: continental India, Gangetic Plains and Burma.

Dhar and Kachroo (1983 a) provided the following additional information for Kashmir Himalaya (endemism above 50 per cent):

Family	total species	endemism %	Genera	total species	endemism %
Adoxaceae	01	100	*Taraxacum*	59	90
Balsaminaceae	14	065	*Swertia*	13	25
Berberidaceae	18	084	*Saussurea*	31	62
Fumariaceae	25	055	*Primula*	21	61
Violaceae	12	054	*Corydalis*	52	55
Gentianacea	55	054	*Pedicularis*	27	52
Saxifragaceae	28	053			
Celastraceae	02	050	*Impatiens*	14	51
Aceraceae	06	030	*Arenaria*	10	51
Valerianaceae	08	050			
Dipsacaceae	06	050			

Table X. Endemism in flora of Afghanistan in relation to that of Kashmir.

Family	Afghanistan (Wandelbo 1966)		Kashmir (Dhar & Kachroo 1983 a)	
	total species	endemic	total species	endemic
Plumbaginaceae	93	75	1	00
Compositae	400	40	334	52
Leguminosae	300	40	125	49
Labiatae	160	35	82	34
Primulaceae	34	35	43	47
Campanulaceae	15	30	12	50
Umbelliferae	140	30	62	54
Boraginaceae	120	30	67	35
Caryophyllaceae	110	30	72	19
Scrophulariaceae	75	30	85	52
Cruciferae	170	20	97	56
Liliaceae	110	18	46	17
Rubiaceae	40	15	21	20
Orobanchaceae	23	13	08	25
Onagraceae	14	07	23	25
GENERA				
Dionysis	10	50	0	00
Acanthophyllum	12	83	0	00
Cousinia	100	80	5	20
Acantholimon	41	75	1	00
Mattiastrum	10	70	3	30
Astragalus	150	70	47	52
Scrophularia	17	53	8	50
Ferula	22	45	2	00
Erysimum	11	36	7	00
Allium	45	33	19	30
Nepeta	40	30	28	59
Rubia	07	29	7	00
Orobanche	19	16	8	47
Campanula	11	15	21	25
Epilobium	13	08	21	05
Bromus	17	00	10	na

Table XI. Endemic situation in genera common to Tien Shan (cf Pavlov 1970) and Kashmir Himalaya

Genus	Western Tien Shan total species	Western Tien Shan endemic species	Kashmir Himalaya total sps	Kashmir Himalaya endemic sps
Aconitum	04	01	?	01
Ranunculus	34	03	27	07
Arabis	10	03	09	02
Draba	19	01	18	05
Viola	12	01	12	06
Arenaria	06	01	10	03
Dianthus	10	04	07	02
Chesneya	02	01	03	01
Hedysarum	22	13	06	04
Oxytropis	48	21	14	06
Bupleurum	06	01	17	07
Salinum	04	04	06	04
Trachydium	02	02	01	01
Lonicera	13	01	18	05
Senecio	08	02	13	04
Cousinia	52	29	05	01
Jurinea	14	07	04	03
Saussurea	09	04	31	17
Scorzonera	13	01	03	01
Taraxacum	27	01	59	51
Androsace	09	01	14	05
Swertia	05	01	13	08
Cynoglossum	03	01	07	04
Euphrasia	05	01	12	08
Pedicularis	14	04	27	13
Scrophularia	11	02	08	03
Nepeta	13	02	28	12
Phlomis	10	06	03	01
Salvia	07	02	08	03
Polygonum	36	04	41	08
Euphorbia	24	03	14	04
Allium	54	11	19	06
Gagea	21	03	?4	01

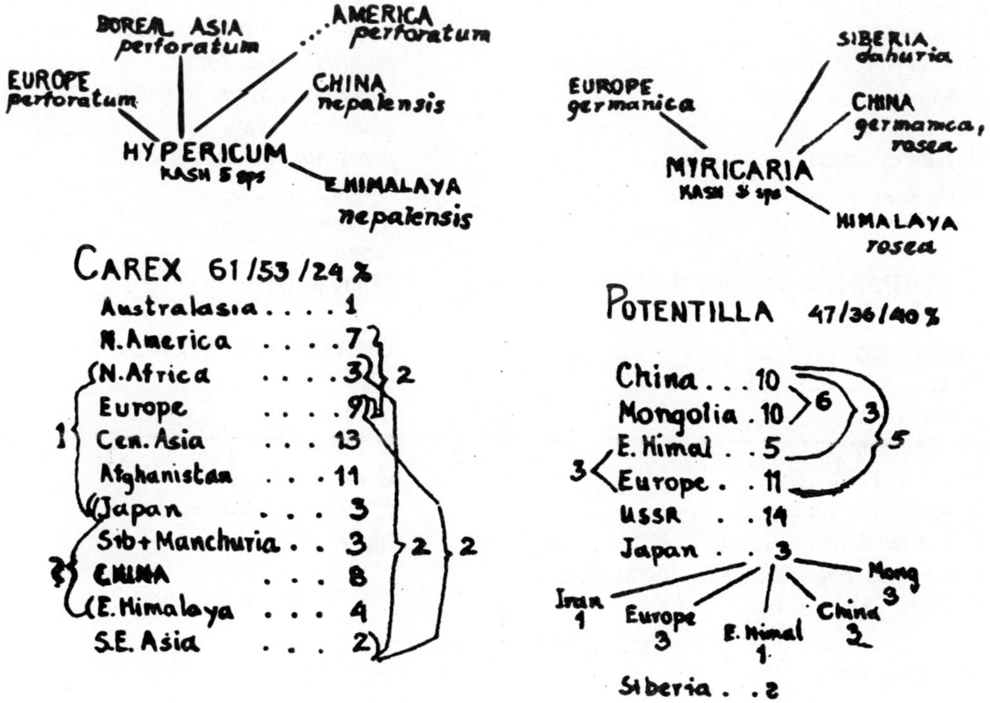

Fig. 62 Phytogeographic affinities of *Hypericum*, *Myricaria*, *Carex* and *Potentilla*.

NEPETA 28/25/50 %

E Him	nil
Europe	nil
Afghanistan	7
Siberia-Manchuria	1
C Asia	2
Iran	1
W Him & China	1
Japan	nil
endemics	13

PRIMULA 21/18/61 %

Afghanistan	5
Tibet	8
Bhutan	2
Khasiamtsm	1
Java	1
Japan	nil
E Him & China	3

denticulata : Afgh, Him, Khasia Sib, N. Burma
prolifera : Khasia, Java.

SPECIES IN TIBET

farinosa . . C Asia, arctic Amer.
sibirica . . circumpolar
tibetica . . Kumaon, E Him.
stuarti . . Afgh, Him.

GENTIANA 35/31/53 %

widespread species :
algida : Eur, boreal Asia, Amer, Jap.
aquatica : boreal & arctic N Amer.
humilis : Asia, Amer.
pseudoaquatica : Jap, E Alps, Manchuria.
squarrosa : Jap.
prostrata : boreal Australia

China	8
E Him	none
Japan	3
China & Jap	nil

CORYDALIS 25/24/55 %

Europe	nil
E Him	4
C Asia	4
E Him & C Asia	2
Japan	nil

Fig. 63 Phyogeographic affinities of *Primula*, *Gentiana*, *Nepeta* and *Corydalis*.

lean towards Central Asia; the presence in Siberia is significant. *Corydalis* does not reach China or Japan. *Primula* reaches Java, perhaps through Bhutan and Khasia. *Potentilla* leans both towards Sino-Japanese-Himalayan provinces as well as westwards. *Mosla dianthera* (Hamilt) Maxim is a similar taxon occurring in Vietnam, the Philippines, Sumatra, Manchuria, Korea, etc.

In *Codonopsis* a true Sino-Japanese element none of our taxa reach beyond Afghanistan and central Himalaya. However our Kobre as show Sino-Himalayan affinity ranging from Central Asia (1 sps in Iran) to E Himalaya. *Epilobium* (21/16/25 %) shows the following affinties:

2 species . . widespread, boreal, circumpolar.
2 . . . . . . widespread, Europe
5 . . . . . . China
2 . . . . . . E. Himalaya
2 . . . . . . Ch- EH
5 . . . . . . endemic

Lonicera (18/8/ 24 %) ranges into Afghanistan-W China and shows the folowing dynamics:

11 species . . . Afghanistan
7 . . . W China
4 . . . E Himalaya
1 . . . Afg-EH
1 . . . Afg-EH- W Ch
2 . . EH- WCh

Saxifraga (Fig. 32) has no relatives in Japan though it reaches E Himalaya and W China; the extensive extension into Siberia and Alaska is sinister. Two species are circumpolar. The widespread elements are the following:

S. cernua L . . Eur, NAmer.
oppositifolia L . . Eur, N Amer.
hircula L . . arctic montane, cricumpolar, south of CA belt.
flagellaris Willd.
ssp *mucronulata* (Royle) Engl et Irn . . endemic to Kash-Pak.
ssp *crassiflagellata* Hulten . . endemic to Kash-Pak.
ssp *stenophylla* (Royle) Hulten . . circumpolar, rare in Sib and Alaska.
ssp *komorovii* (Lov-Lozin) *Hulten* . . Tien Shan
diversifolia Wall ex Seringe . . EH, Assam, N Burma, W Ch.
moorcroftiana Wall ex Sternb . . EH, S Tb, W Ch.

The species common to Central Asia and (western) China may be grouped into two:

(A) SPECIES RANGING INTO NW HIMALAYA

Acantholimon lycopodioides Boiss, *Anaphalis virgata* Thoms, *Bupleurum gracillimum* Klotsch (only Kar-NWH), *Carex melantha* Cam var *moorcroftii* Kuk, *C.tristis* MB (Iran, Turk), *Chesnea cuneata* (Bth)Kom, *Codonopsis clematidea* Cl, *Crepis flexuosa* Cl, *Dracocephalum stamineum* K et K, *Elymus dasystachys* Trin, *Inula obtusifolia* Kem var *clarkei* Hk. f., *Lactuca tatarica* CAM, *Lonicera semenovii* Rgl, *Melilotus officinalis* Lam, *Myricaria elegans* Royle, *Oxytropis humifusa* K et K, *Rubia tibetica* Hk. f., *Stachys tibetica* Vatke.

(B) SPECIES RANGING INTO HIMALAYA

Aster flaccidus Bge ssp *flaccidus* Grierson (Sib), *Bupleurum falcatum* L var *marginatum* (Wall.) CB Cl, *Draba oreades* Schrenk, *D. stenocarpa* Hk. f. et T (Turk), *Elymus nutans* Griseb, *Sibbaldia cuneata* Hornem, *Waldheimia tridactylites* K et K.

Our Borages do not range into west China or Japan, 44% are common with USSR as a whole, 27% with Central Asia and just 2% with eastern Himalaya.

Of the other elements not in China and Japan may be mentioned the following: some of these are common to adjoining areas also:

Anaphalis contorta (Don) Hk. f. . EH, Afg.
Carex borii Nelmes . . WH, Kar.
nivalis Bott . . H
pamirensis CB Cl ex F Fedt. WH
pseudofoetida Kuk Tb
sibirica . . Tb, NWH.
Dactylorhiza hatagirea (Don) Scop . . H
Kobresia capillifolia CB Cl . . WH
macrantha Boeck . . Iran, WH
nitens CB Cl . . Kar, WH
schoenoides Boeck . . Iran, Sib, H.

However *Lycium ruthenicum* Muer occurs in W China, but is not reported from the Himalaya!

Among the taxa that reach western China and even Taiwan but do not range int Japan are the following:

Anemone alnoides D Don, *Astilbe rivularis* Hamilt, *Betula alnoides* D Don, *Callicarpa monophylla* Vahl, *Cardamine macrophylla* Willd. var *venusta* OES (N Korea), *Cimicifuga foetida* L (Sib, Korea), *Clematis buchanianus* DC, *C. connata* DC, *C. montana* Hamilt ex DC, *Hackelia uncinata* (Royle ex Bth) OEC var *brachytuba* (Diels) Hara, *Fraxinus floribunda* Wall ex Royle, *Meliosma pungens* (Wall) Walpers. *Neolitzea umbrosa* (Nees) Gamble, *Perilla frutescens* (L) Britt., *Sedum multicaule* Wall ex Lindl. *S. wallichianum* Hook.

The elements common to western China and parts of N and W Asia are the following :

Aconitum rotundifolium K et K, *Chorispora sibirica* DC (W Sib), *Cotoneaster uniflora* Bge, *Delphinium brunonianum* Royle, *Draba altaica* (Lam.) Bge (H), *Eritrichum rupestre* (Pallas) Bge var *pectinatum* (Pallas) Brand, *Malcomia africana* R Br (Turk), *Parnassia ovata Ldb, Paraquilegia uniflora* Drum et Hutch, *Sibbaldia tetrandra* Bge, *Stellaria graminea* L (Mong, E Sib, Eur).

None of our Ranunculi reach Japan. It would be worthwhile to examine relationships of the local taxa with *R. pulchellus* CAM var *stracheyanyus* Hand.-Mazzt (CA, Sib, H, Tb, Mong, W & N Ch). Our *Philadelphus (tomentosus* Wall ex G Don) has no relatives in China or Japan. *Cerastium pusillum* Ser (W Ch, N Mong) does not occur in Japan.

Teucrium quadrifolium Hamilt ocurs in Kashmir-Sikkim-Khasia Hills, N Burma, Thailand and China, but *T. viscidum* Blume occurs in the Himalaya, W & S China, Taiwan, Korea, Japan and Malaysia, rather interesting! Similarly *Senecio chyrsanthemoides* DC ranges into W China and its "relative" *S. scandens* Hamilt extends further the range of the genus into the Philippines, China and Japan. There is need for taxonomic revisions of such genera.

Our Loniceras range into Sikkim (*L. angustifolia* Wall ex DC, *L. obovata* Royle ex Hk.) but *japonica* Thunb occurs in China, Korea, Japan. We need to study relationships of these taxa.

Among the taxa that reach Central America, *Stellaria* is typical temperate element, showing the following range:

S. alsine Grimm . . Eur, CA, Mediterranean, Scandinavia, Asia Minor, N Asia, Ch, EH.
graminea L . . WS, Eur, Dzungaria-Kashgar, Mong, N Asia.
media (L.)Vill . . circumpolar, WS. Eurasia, C. Amer.

"A close revision of *Stellaria* from north India to east Siberia is strongly needed" (Hara, 1966).

S. media L. sensu Edgw (FBI 1,230 pp) is *S. media* var *procera* Richter and distributed in Europe, N Africa, N Amer (Hara, 1966) whereas its var *media* Edgw et Hk is cosmopolitan.

Vaccaria pyramidata Medik is circumpolar: Eurasia, Australia, Japan, Central Asia. Genera like *Myricaria* illustrate W-N-E extensions of some of the local elements (Fig 62).

M. germanica ranges into Europe; Dahuria to Siberia, into China and into the Himalaya.

M. germanica Desv var *prostrata* (Bth ex Hk f.) Dyer occurs in Kumaon and Sikkim!

Hypericum connects Europe, boreal Asia, America and the Sino- Himalaya region: *H. perforatum* occurs in boreal Asia, Europe and America, *nepalensis* in China and E Himalaya (Fig. 62).

Our Sedums common to Central Asia do not extend beyond western Himalaya; nor do some species of *Sempervivum* and *Ribes*!

S. adenotrichum Wall ex Edgw . . Afg-Kash-Nepal
crassipes Hk. f. et T . . Kar, H, Tb, W Ch ?.
ewersii Ldb . . W Mong, Altai, W Ch, Nepal.
fastigiatum Hk. f. . . Kr, ? H.
heterodontum Hk. f. et T . . WCh, NWH.
acuminatum Dcne . . Kar, NWH.
Ribes orientalis Desf. . Caucasica, W Ch, Mong.

Moneses uniflora (L) AGray is yet another circumpolar species but lacks east-west races (Hulten, 1971). *Sanicula elata* Buch-Ham ex D Don is widespread in east and west and *Goodyera repens* is widespread locally between the mountains of Europe and Kashmir(Figs. 33,54).

Our Pediculari common to Central Asia and China show the following local range (Fig.27).

WH . . . *P. albida* Penn, *bicornis* Klotz.
NWH. . . *P. brevirostris* Penn., *cederi* Vahl ssp *typica* Penn, *pectinata* Wall. ssp *typica* Penn.
Tb . . . *P. albida, oederi* ssp *typica.*

Angelica another discontinuous element of north temperate zone and Australasia is represented by *A. glauca* Edgew alone and is endemic to western Himalaya.

Concluding remarks

Affinities with the Antartic (Antartic Continent) are revealed by discontinuous families and genera, eg Hippuridaceae an element of north and South America, Valerianaceae of the same area and S Africa and Malaysia; Dipsacaceae and Tamaricaceae of Eurasia and Africa; Papaveraceae of north temperate N America-Andes-S Africa-Australia; and Orobanchaceae of S America-Australia and north temperate zone. Admittedly these disjunctions present the relics of former wide continuous distribution patterns, the intervening areas having been populated sometimes and now covered by seas. I have no proof of any of these taxa arising independently in separate areas of distribution, thus exhibiting evolutionary parallelism or convergence. This needs more search.

The local elements of these taxa range as follows:

Hipuris vulgaris L., WS in northern hemisphere.

Valeriana (8 sps, Figs.30,33): *hardwickii* Wall . . Malaya, Java, EH, WCh
dioica L . . WS in N Asia, Eur.
endemics . . 4 sps.

Dipsacus (2 sps) . . H

Morina (3 sps) . . Eurasia, 1 WS in Mediterranean, Afg.

Scabiosa (1 sps) . endemic Kash-Pak.

Hypecoum (2 sps) . Eurasia, reaches Ch, Jap.

Meconopsis (1 sps) . . endemic (genus, Eur, Sino-Himalayan).

Papaver (2 sps) . 1 reaches Sib, Mong, Ch.

Lathrea (1 sps) . . Central Europe.

Orobanche (7 sps) . . 1 WS in Eur-Amer, Caucaus, boreal Asia.

Myricaria (3 sps) . . Eur, Ch, H, Sib.

Tamarix (2 sps) . . WS, Eur.

Perhaps we could also mention here two taxa predominantly tropical: Araliaceae represented by *Aralia cachimiricum* Decne (Afg-WH-EH) and Acanthaceae (*Strobilanthes* 4 sps) of Magasay-Australia-Pacific Islands. Our elements range into Afghanistan-East Himalaya.

Veronica (26/19/28 %) an element of north temperate zone and Australasia has none of its local elements ranging into Africa. *Astragalus frigidus* (L.)A Gray reaches W C and N Amer; *A. gracilipes* Bth & Bge and *A. ophiocarpus* Bth & Bge reach Baluchistan; so does *Aquilegia vulgaris* ssp *moorcroftiana* (Wall.) Bruhl etc. Species of *Silene* which reach N Afr and of our Festucas, *F. rubra rubra* is widespread and *F. kashmiriana* is restricted in distribution:

Festuca altaica (Hack)Roshev . . Alatau, H, P (not recorded from Kashmir).
kashmiriana Stapf . P, WTb, Kara.
ovina L ssp *coelestis* St-Yves var *coelestis* Mgh-Dbg . . Tien Shan, P-A, WTB, WCH, Kar (not reported from Kashmir)
rubra L. ssp *rubra* Hack forma *barbata* (Hack) R. Lit. . . . Eurasia, Tb, NAmer, Australia.

Among the species common to west China, Himalaya and Malayasia may be mentioned : *Myrica esculenta* Hamilt ex D Don, *Persicaria posumbu* (D Don) H Cross, *Rubus niveus* Thunb var *rosaefolius* (Hk. f.) Hara, *Indigofera trifoliata* L (also ranging into N. Australia!).

The genus *Silene* shows interesting links with (a) Iran-W Ch, as well as (b) Japan:

S. conoidea L . . Turk , Iran, W Ch.
gonosperma (Rupr) Bocq . . Kar, H, WTb.
moorcroftiana Wall . . Hind, Kara, W Tb, WH.
tenuis Willd . . Sib, W Mong, W Ch, Kar, Hind, WH.
vulgaris (Moench) Garcke . Eurasia, N Afr, Jap, N & S Amer.

Before concluding I might refer to Takhtajan's (1969) floristic regions of the world, which he divided into six kingdoms. I may recall here that he followed largely Good in his treatment. Of these the first Holarctic Kingdom is more relevant here (Fig. 64). This he divided as follows: (only relevant items mentioned here).

Floristic regions of World: six kingdoms.

1. Holarctic . . A) Boreal subkingdom
   i) arctic and subarctic region
   ii) Euro-Siberian
   iii) Eastern Asian
   iv) Atlantic N. American
   v) Rocky mts.

Fig. 64. Floristic regions of the world (part shown) cf Takhtajan 1969).

B) Tethyan (A M) subkingdom

vi) Macaronesian

vii) Mediterranean

viii) Irano-Turanian

C) Madrean (ix)

2. Palaeotropical kingdom

A) African subkingdom . . x) Saharo-Sindian region.

C) Indo-Malesian subkingdom. . xvi) Indian region.

The Regions 10 and 16 may have some historical significance (if any) to our area.

Zohary's (1971) treatment of Middle East is also interesting. It is enumerated below:

A. The Euxine-Hyrcanian territory:

mountains of north Turkey and northern slopes, in Late Tertiary of Elbruz mts, a sub-province of Pontic Province, in Late Tertiary and Pleistocene it served as a refugium for many plants which were driven southward by intense cold.

B. The Mediterranean territory:

narrow belt along the Mediterranean Sea never an important centre of speciation, endemics of a diffuse type. It is divided into two phytochoria; the Anatolian (vegetation more mesic) and Palestinian.

C. The Irano-Turanian territory:

occupies half of the surface of Middle East and most of Egypt and Arabia, extending to Palestine, Syria, Iraq, east Turkey and Iran. The following districts/enclaves were recognised here: (1) the Mesopotamian 2) the Irano-Anatolian. 3) Armeno-Anatolian. 4) the Kurdo-Zagrosian 5) Irano-Armenian enclave and 6) the Turanian enclave.

D. The Saharo-Arabian territory:

a poor floristic stock of its own, mainly an admixture of B and C and E territories.

E. The Sudanian territory:

a hot ancient desert flora, rich in endemics.

The Irano-Turanian territory assumes importance and Zohary considered it an important centre of differentiation and speciation. "One of the most ancient desert and steppe, pre-Mediterranean floras" had much wider areas and its "relic stations serve as milestones of former migration tracks from the main region to north Africa and southern Eurasia. For our affinities of this territory we may refer to Table V again.

Now we may consider western Himalaya in the light of remarks of Meusel (1971) who stated that in this area "there exists gradation from semi-humid monsoon climate to a typical semiarid etesianclimate. Therefore we can meet both outliers of the humid Sino-Japanese flora . . and members of the Mediterranean flora". He considered the flora of the Himalaya as out-post of the Mediterranean (Tethys) flora. He also believed that the "Himalaya is a bridge between Sino-Japanese and the sub-Mediterranean-central European .

His assessment with respect to western Himalaya appears to be plausible in p. max. p.; but like Takhtajan he did not recognise NW and W Himalaya as distinct outliers, nor did he consider the impact of Sino- Himalayan province on western Himalaya. The northwestern Himalaya of which Kashmir is a part does embrace his semi-arid etesian climate though arid regions need consideration. In Kashmir Himalaya we have scarce Sino-Japanese elements, respectable Sino-Himalayan elements and in relation to Mediterranean we lean more heavily towards Central Asia, the latter is said to be rich in Irano-Turanian elements. In fact it is a major part of the Irano-Turanian. But that is not the whole situation!

The interesting floristic panorama of Kashmir discussed so far leads to the conclusion that it owes its parentage (*nec origin*!) to the Tethys Sea Region or what is also called the AMR. From the AMR (Fig. 65) we derive two principal floral provinces: the Mediterranean (s str) and the Asia Mediae; the former would represent western part of both the shores of the present Mediterranean including parts of southern Europe; to the latter would belong what largely is the Irano- Turanian region and what I consider Asia Mediae proper. From AM (s str) are derived Asia Centrali and west Himalayan region; in other words I draw a line between NW and W Himalaya in the same way as between W Asia Mediae and the rest of it. I couple CentralAsia-W China and Central Asia-Kashmir as subregions of Central Asian Region. To the west Himalayan region I allot Tibet-west Himalaya and W China-Himalaya subregions. I have also attempted to draw a line between Sino- Himalayan and Japano-Himalayan regions of the principal Sino-Japanese Region; and consider the former to be related to our area coupling W Chinese- Kashmir subregion with Tibet-North Central China subregion; the former would approach Asia Centrali-Kashmir subregions. These presumptions might need further support!

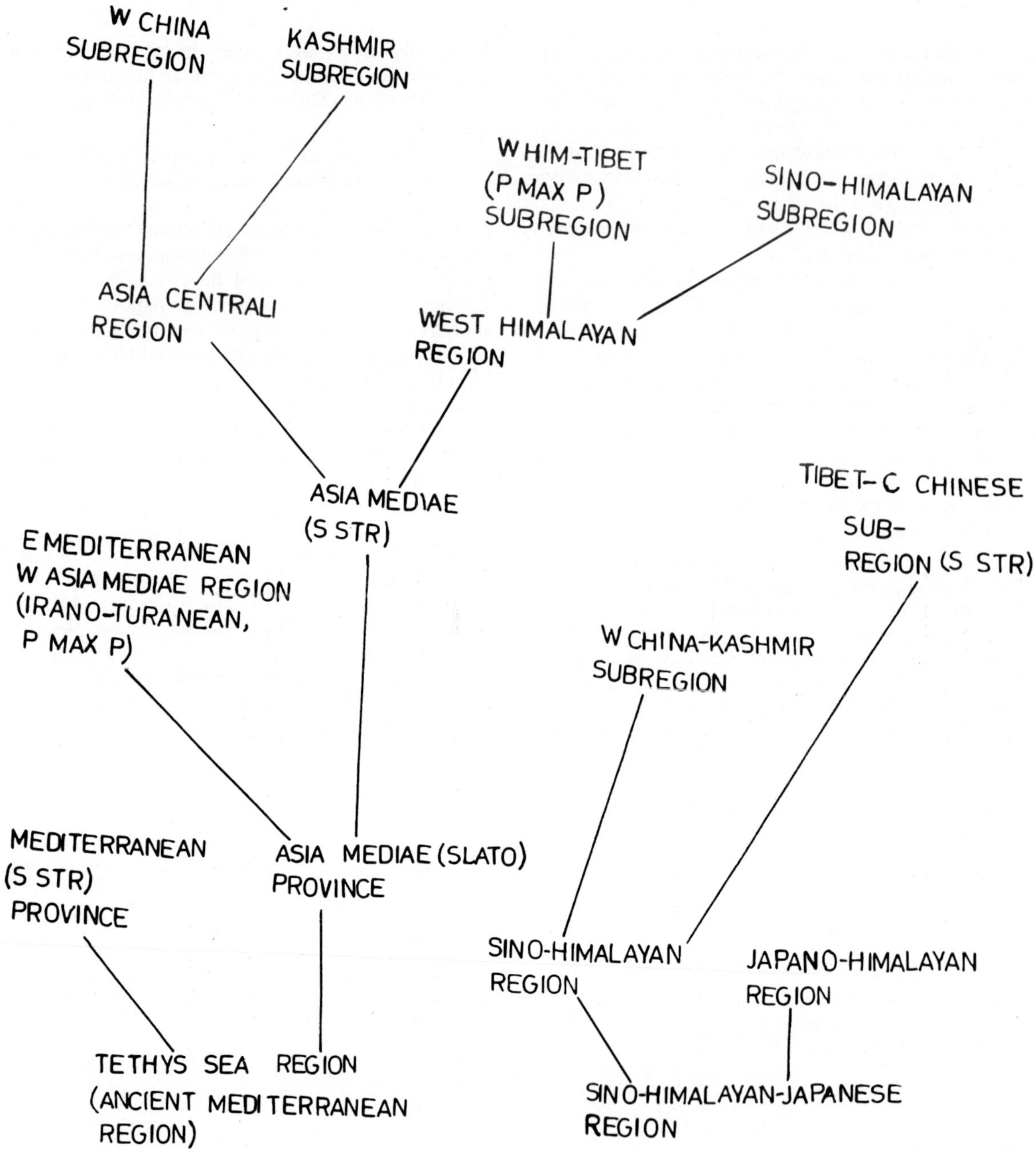

Fig. 65 Probable relationships of various floral provinces of Kashmir and adjoining regions.

APPENDIX

Appendix I. Distribution of Pteridophytes in Soviet Central Asia and Kashmir (common species[#]) (+, present ; - absent)

Species	Uzbekistan	Tadjikstan	Kirgiz	Kazakhstan	Kashmir
1. Adiantum capillus-veneris L.	+	+	+	+	+
2. Asplenium adiantum-nigrum L.	-	-	+	-	+
3. [†]A. haussknechtii Godet et Reut.	-	+	-	-	-
4. A. pseudofontanum Koss	-	+	-	+	+
5. A. rutamuraria L.	+	+	+	+	+
6. A. septentrionale (L.) Hoffm	-	-	-	+	+
7. A. trichomannes L.	+	+	+	+	+
8. A. viride Huds.	+	+	+	+	+
9. Athyrium filix-femina (L.) Roth	-	-	-	+	+
10. Botrychium officinarum (L.) Sw	+	+	-	+	+
11. B. multifidum (Gmel) Rupr	-	-	-	+	-
12. Ceterach officinarum willd	+	+	+	+	+
13. Cheilanthes persica (Bory) Mett et Kuhn	+	+	+	+	+
14. Cystopteris fragilis (L.) Bernh. (=C. filix-fragilis (L.) Borbas)	+	+	+	+	+
15. Diphasianstrum alpinum (L.) Holub	-	-	-	+	+
16. D. complanatum (L.) Holub (=Lycopodium alpinum L.,. L. anceps Wallr)	-	-	-	+	-
17. Dryoptersis filix-mas (L.) Schott	+	+	+	+	-
18. D. fragrans (L.) Schott	-	-	-	+	-
19. D. barbigera (HK) O. Kuntze (=D. komarovii Koss)	+	+	+	+	+
20. D. mindschelkensis Pavl	-	-	-	+	-
21. D. carthusiana (Vill) HP Funchs (= D. spinulosa (Sw) Watt)	-	-	-	+	-
22. Equisetum arvense L.	+	+	+	+	+
23. E. fluviatile L. (=E. heleocharis Ehrb.)	-	-	-	+	-
24. E. hyemale L.	-	-	+	+	-

Species	Uzbekistan	Tadjikstan	Kirgiz	Kazakhstan	Kashmir
25. E. palustre L.	-	-	+	+	+
26. E. pratense L.	-	-	+	+	+
27. E. ramosissimum Desf	+	+	+	+	+
28. E. scirpoides Mixch	-	-	-	+	-
29. E. sylvaticum L.	-	-	-	+	-
30. Gymnocarpium dryopteris (L.) Newm	-	-	-	+	-
31. G. robertianum (Hoffm) Newm	-	-	-	+	-
32. G. fedtschenkoanum Pojark	-	+	-	-	-
33. Huperzia selago (L.) Bernh Ex Schrank et Mart (= Lycopodium selago L.)	-	-	-	+	+
34. Lepisorus clathratus (Cl.) Ching	-	-	-	+	-
35. Lepisorus thunbergianus (Kaulf) Ching (=Lycopodium alberti Rgl, L. lineare Tnunb).	-	-	+	-	-
36. Lycopodium annotinum L.	-	-	-	+	-
37. L. clavatum L.	-	-	-	+	-
38. Metteuccia struthiopteris (L.) Rtod (= Struthiopteris filicastrum All.)	-	-	-	+	-
39. Marsilea aegyptiaca Willd.	-	-	-	+	-
40. M. quadrifolia L.	+	+	-	+	+
41. M. strigosa Willd.	-	-	-	+	-
42. Ophioglossum bucharicum (O et B Fedt.) O. et B. Fedt	+	+	-	-	-
43. O. vulgatum L.	+	-	+	-	-
44. Phegopteris connectolis (Michx) Wall (=Thelypteris phegopteris (L.) Bory ex Newm)	-	-	-	+	+
45. Polypodium vulgare L.	-	-	+	+	-
46. Polystichum braunii (Spenn) Fee (=P. lobatum (Sw) Presl).	-	-	-	+	-
47. P. aculeatum (L.) Roth (=P. lobatum (Huds) Bast)	-	+	+	-	-
48. P. lonchitis (L.) Roth	-	-	-	+	+
49. Pteridium aquilinum (L.) Kuhn	-	-	-	+	+

Species	Uzbekistan	Tadjikstan	Kirgiz	Kazakhstan	Kashmir
50. Salvinia natans (L.) All	+	+	+	+	+
51. Selaginella aitchisonii Hiern	-	-	+	-	-
52. Thelypteris palustris Scott(= Dryopteris thelypteris (L.) A Gray)	-	+	+	+	+
53. Woodsia alpina (Botton) Gray)	-	-	-	+	+
54. W. ilvensis (L.) R Br	-	-	-	+	-

† = Asplenium samarkandense Koss
* = Ophioglossum vulgatum L. var bucharicum O et B Fedt.
# = These are the only species mentioned in the floras of the respective areas in Central Asia

REFERENCES

Agrawal D 1984 *Cenozoic changes in Kashmir-the multidisciplinary data*. 1-12*

Aitchison J E T 1881-82 On the flora of the Kurram valley. J Linn Soc Bot 18, 1-113; 19, 139-200.

____1888 The botany of the Afghan Delimitation Commission. *Trans Linn Soc Bot* 2 Ser. 3, 139.

Allchin Bridget and Allchin Raymond 1968 *The Birth of Indian Civilization*. Penguin books.

Avetisyan V E 1974 Comparison of the alpine crucifers of the Caucasus and Central Asia. *Biol Zh Arm* 27(1), 99-100.

Axelrod D I 1958 Evolution of the Madro Tertiary Geoflora. *Bot Rev* 24, 433-509.

Basu Arabinda and Ananda Pal 1980 Human remains from Burzahom. *Anthropol Surv India Calcutta*.

Bhargava P D 1971 *India in Vedic Age*. Lucknow.

Bhatt D K 1976 Stratigraphical status of the karewa group of Kashmir Indian Himalayan Geolog 6, 197-208.

Breckle Siegmar W 1974 Notes on alpine and nival flora of the Hindukush, east Afghanistan. *Bot Not* 127, 278-284.

Burbank D W and Johnson G D 1982 Inermountane-basin development in the past 4 my in the northwest Himalaya. *Nature* 298, 432-436.

Burkill I H 1909 *A working list of the flowering plants of Beluchistan*. Suptd Govt Printing India Calcutta 1-136.

Buth G M and Kaw R N 1984 *Plant husbandary in neolithic Burzahom Kashmir* 109-113.

Bisht R S and Gaur G S 1982 Investigation of palaeobotanical remains from Semthan Kashmir *Man & Env* 6, 41-45.

Chaney R W 1947 Tertiary centres and migration routes. *Ecol Monog* 17, 139-148.

Chatterjee D 1939 Studies on the endemic flora of India and Burma. J Asiatic Soc Bengal 5, 69.

[†]Coles J M and Higgs S S 1969 *The Archaeology of Early Man*. London.

Contandriopoulos J 1962 *Researches sur la flore endemique de la Corse et Sur ses origines. These Montpallier*.

Crawford A R 1974 The Indus suture line, the Himalaya, Tibet and Gondwanaland *Geol Mag* 111, 369-480.

Czerepanov S K 1979 *Definitorium Plantarum Vascularium Badachschaniae*. Nauka Leningrad.

____1973 Additamenta et Corrigenda ad 'Flora URSS' (Vols 1-30). Hayka.

____1981 *Plantae Vasculorea URSS*. Nauka Leningard.

Das S K 1944 *The Economic History of Ancient India*. Calcutta.

Davis Richard S 1984 *Pleistocene climatic fluctuations and palaeolithic settlement distributions in Soviet Central Asia and surrounding territories*. 215-227*.

______Ranov V A and Dodonov A E 1980 Early man in Central Asia. *Scient Amer* 243 (6), 130-137.

de Terra H and Paterson T T 1939 *Studies in the Ice Age in India and associated human culture*. Publn no 493 Carnegie Institute Washington DC.

Devey J and Bird J M 1970 Mountain belts and the new global tectonics. *J geophys Res* 75, 2625-2647.

Dhar U and Kachroo P 1983 (a) *Alpine Flora of Kashmir Himalaya*. Scientific Publishers, Jodhpur, India, pp 1-213.

______and ____1983 (b) Some remarkable features of endemism in Kashmir Himalaya IN An Assessment of Threatened Plants of India (ed S K Jain). Bot Surv India Howrah.

Deanelli G 1922 Spedizione italiana de Filippi nell 'Himalaya Caracorum e Turchestan Cinese (1913-1914) ser 2 Risultati geologicie geografici 3 Studi Sui giaciale Zanichelli Bologna.

Dikshit K N 1982 The neolithic cultural frontiers of Kashmir. *Man and Env* 6, 30-36.

Dobremez Jean-Francois 1972 Les grands divisions phytogeographiques du Nepal et de Himalaya (Eng summ). *Bull Soc bot Fr* 119, 111-120.

Dodonov A E 1984 Neogene-quaternary boundary in Central Asia and inter-regional correlation. 197-203.

______and Penkov A V 1977 some results of stratigraphy on watershed loesses of Tadjik depression. *Bull Study Quatern Rev* 47, 67-76.

_____and Lomonov S P 1984 *Stratigraphy and podogenesis of loess formation of southern Tajikstan.* 223-225.*

Duthie J F 1898 The botany of the Chitral Expedition. *Rec bot Surv India* 1(9), 139-181.

Engler A 1879 Versuche einer Entwicklungsgeschichte der pflanzenwelt inbesondere der florengebiete seit dar Teria Periode. Leipzig pp 202. 1899 Die entwicklung der pflanzengeographic in den letzten hundert Jahren A V Humbold- Centenarschrieft. Berlin.

Favarger C 1972 Endemism in the mountain floras of Europe IN *Taxonomy, Phytogeography and Evolution* (ed D H Valentine). Academic Press London, & New York 191-204.

Conspectus Florae Asiae Mediae 1-7 1968-1983 (Vols 8,9 in press.)

Plantae Asiae Central, 1-6 1965-71. Acad Scient URSS Moscow/Leningrad.

Flora of British India 1-7 1875-1897 London.

Flora Khazakastanica 1-9 1959-1966, Acad Scient Kazak SSR Alma Ata.

Flora Kirigiz CCP 1-11 suppl 1,2, 1952-1970, *Acad Scient Kirgiz* SSR Tashkent.

Flora of Pamir, d'apresles explorations personnelles en 1901 et celles des voyages pre'cedents (by Mm Olga Fedtschenko) *Acta Horti Petropolitani* 21(3), 235-471, 1903. (Printed seprately by Imperial Soc of Natls Moscow 1-250, 1903. suppls 1-4 *Acta Horti Petropolitani* 24(2) 125-260, 1904; 24(3) 315-354, 1905; 24, 99-126, 1907; 24, 455-4 514, 1909).

Flora of Tajik SSR 1-7 1957-1984 (vols 8,9 in press). Acad Scient Tajik SSR Dushanbe.

Flora Turkmenii 3,6 1945-1948. Turk Acad Sci-CCCP, Ashkabad.

Flora Uzbekistania 1-6 1941-1962. Uzbekistanicae Acad Scient URSS Tashkent.

Florin R 1963 The distribution of conifers and cycads in time and space. *Acta Hort Berg* 20, 121-312.

Freitag H 1971 Die Naturliche Vegetation Afghanistans. Vegetation 22,285-344.

_____1977 The pleniglacial, late glacial and early Postglacial vegetations of Zeribar and their present day counterparts. *Palaeohistoria* 19, 87-95.

Ganesser A 1964 *Geology of the Himalayas.* Inerscience London.

Garsen M S and Krs M 1976 Geophysical and geological evidence of the relationships of Red Sea transverse tectonics to ancient fractures *Geol. p Soc Am Bull* 87, 169-181.

Gupta H P Sharma Chhaya Dodia Rekha Mandavia Chetna and Vora A B 1984 *Palynostratigraphy and palaeoenvironments of Kashmir* Hirpur loc 3, 75-90.*

Haeckel J and Troll W 1938 Botanische Ergebnisse der Deutschen Hindukush Expedition 1935 *Feddes Repert Beih* 108, 1-72.

Hara H 1966, 1971 The Flora of Eastern Himalaya 1,2 Univ of Tokyo Press.

Hartman Hans 1966 Beitragezur Kenntnis der Flora des Karakoram. *Bot Jh* 85 (2,3) 259-323; 324-406.

______1983. Pflanzenges ellschaften entlang der Kashmir route in Ladakh. *Jahrb var Schutz Bergwelt* 48, 131-173.

______1984 Neue und wenig bekannte Bluetenpflanzen aus Ladakh miteinem Nachtrag zur Flora des Karakoram. *Candollea* 39, 507-537.

Hawkes Chr Hawkes J and de Terra P 1934 Palaeolithic human industries in the northwest Panjab, Kashmir and their geological significance. *Mem Conn Acad Art & Sci* 8, 1-14.

Hemsley W B 1902 The Flora of Tibet and High Asia *J Linn Soc Bot* 35(244), 124-265.

Hulten E 1962 *Circumpolar Plants* 1(1962), 2(1972). Almiqvist & Wiksell Stockholm

Ikonnikov Sergey 1963 *The Definitorium Plantarum Pamiricum.* Dushanbe.

_____1965 Recent data on the flora of Pamir IN Studies of the flora and vegetation of high mountain areas (ed V N Sukachev). IPST Jurusalem 237-242.+

Jha D N 1977 *Ancient India an introductory outline.* Peoples Publishing House New Delhi 134, pl 19.

Joshi R V et al 1974 Quaternary glaciation and palaeolithicsites in Lidder Valley (Jammu and Kashmir). *World Archaeology* 5,369- 379.

Kamelin R V 1965 on the generic endemism of the Middle Asiatic flora. *Bot J* 50 (12), 1702-1710.

______1967 On some remarkable anomalies in the flora of the Middle Asiatic Mountain Province. *Bot J* 52(4), 447-460.

Kanai H 1963 Phytogeographical observations on the Japan- Himalayan elements. *J Fac Sci Univ Tokyo* ser 3 Bot 8(8), 305-339.

Kennedy Kenneth A R 1984 The hominid fossil record in Central Asia-new discoveries and interpretations. 205-213*.

Kitamura S 1960 Flora of Afghanisthan. Kyoto 1,2.

Konnov A. A. 1973 Flora of Juniper societies of Shahristan (on the northern Tajikistan) Dushanbe (in Russ).

Kusumgar Sheela Bhandari N and Agrawal D P 1984 Fission track ages of the Romushi Lower Karewa Kashmir. 245-247.

______and Agrawal D P 1984 chronology of the Kashmir loess. 119-121*.

Lace J H and Hemsley W B 1891 A sketch of vegetation of British Baluchistan with descriptions of new species. *J Linn Soc Bot* Lond 28, 288-327.

Lapyajaeva V S 1976 *The Forest Resources of Pamir-Alai.* Nauka Leningrad.

Li H L 1944 The phytogeographic division of China with respect to Araliaceae. *Proc Acad Nat* Philadelphia 96, 249-277.

Linchovsky I A and Prozorovski A V 1949 The basic principles of the distribution of the vegetation of Afghanistan. *Kew Bull* 1949, 179-214.

Lomonov S P and Ranov V A 1984 The peculiarities of the Pleistocene palaeosol formations and distribution of embeded paleolithic tools. 227-240*.

Mackay Ernest 1976 *Early Indus Civilization* (rev & enlarg by Dorothy Mackay). Indological Book Corporation Delhi.

Meher-Homji V M and Misra K C 1973 Phytogeography of the Indian sub-continent IN *Progress of Plant Ecology in India* (ed R Mishra, B Gopal, K P singh and J Singh). New Delhi 9-90.

Melville R 1981 Vicarious plant distributions and palaeography of the Pacific region IN *Vicariance Biogeography*-a Critique (eds G Nelson & D E Rosen). Columbia Univ Press New York 238-302.

Meusel H 1971 *Mediterranean elements in the flora and vegetation of western Himalaya* IN Plant Life of Southwest Asia (eds P H Davis, P C Harper & I C Hedge). *Bot Soc Edinb* 53-72.

______and Schubert R 1971 Beitrage zur Pflanzengeographie des west Himalaja 1 Die Araltypen 2 Die pflanzengeographische stellung und gleidevung des west Himalajas. *Flora* 160, 137-194; 537-606.

Mani M S 1978 *Ecology and Phytogeography of High Altitude Plants of Northwest Himalayas.* Oxford & IBH Publishing Co New Delhi.

Mitra P 1927 *Prehistoric India.* Calcutta.

Mongit Alexander 1959 *Archaeology in the USSR.* Moscow

Nayar M P 1980 Endemism and patterns of distribution of endemic genera (Angiosperms) in India. *J Econ Taxon Bot* 1, 99-110.

Osborn K 1911 *Man of the Old Stone Age.* London

Ojha K 1968 The *History of Foreign Rule in Ancient India.* Gyan Prahashan Allahabad.

Ovczinnikov P N (Chief ed) 1971 *The Flora and Vegetation of Varsob river Canion.* Nauka Leningrd.

____Sidorenko GT and Kaletkina 1973 *a Vegetation of Pamir-Alay.* Leningrad

___Sidorenko G T & Kaletkina 1973 b *The Report of the Inernational Symposium on Eco-Physiological Foundation of Ecosystem Productivity in Arid Zone* Leningrad-Dushanbe June 7-19 1972 (Dusshanbe 1973).

Pavlov V N 1970 The endemism of the western Tien Shan flora. *Bot J* 55(9) 1232-1242.

Porter Stephen C 1984 *Extent of Late-Pleistocene glaciers in Afghanistan basedon interpretation on landstat imagery.* 191-195.

Ronov V A 1982 The Hissar Neolithic culture of Soviet Central Asia. *Man & Env* 6,63-71.

Rao C K 1972 Angiosperm genera endemic to the Indian floristic region and its neighbouring areas. *Indian Forest* 98, 560-566.

Raven Peter H 1971 The relationships between 'Mediterranean.' floras IN Plant Life of Southwest Asia (eds PH Davis, PC Harper & IC Hedge). 119-134 Edinburg.

____1980 Plant tectonics and southern hemisphere biogeography

Sahni B 1925 The southren fossil floras Pres add Geol Soc in *Proc 13th Indian Sci Congr.* Bombay 229-254.

____1935 (a) The relations of the Indian Gondwana flora with those of Siberia and China. *2nd Congr Carboniferous Stratigraphy Holland Compte Rendu* 1, 517-518 (1937).

____1935 (b) *The Glossopteris flora of India.* Proc 6th intern bot Congr Amesterdam 2,245-247.

____1935 Permocarboniferous life province with respect to India. *Current Sci* 4(6) 385-390.

____1936 The Karewas of Kashmir. *Current Sci* 5(1) 10-16.

____1936 The Himalayan uplift since the advent of Man, its cult-historical significance. *Current Sci* 5(2) 57-61.

____1937 Speculation on the climates of Lower Gondwanas of India. 17th intern geol Congr Abs 217-219 Moscow.

____1939 The relation of Glossopteris flora with Gondwana glaciation. *Proc Indian Acad Sci* 9(1) B,1-6.

Sankalia H D 1963 *Prehistory and Protohistory of India and Pakistan. Bombay*

____1971 New evidence for Early Man in Kashmir. Current Anthropol 12,538-554.

____1977 *Prehistory of India.* New Delhi.

*Sarkar S S 1962 *Ancient Races of Baluchistan, Panjab and Sind.* Calcutta.

Stenbridge Jasper H 1965 *The World-a general regional geography.* Oxford University Press London x,530.

Sun Honglei 1983 Land resources and agricultural utilization in Xizand(Tibet) autonomous region China. *Mountain Res & Develop* 3(2), 143-148.

Takhtajan A 1969 *Flowering Plants,* origin and dispersal. Oliver & Boyd Edinburgh

Troll C 1938 *Der Nanga Parbat als deutscher Forschung Zeitschr Gesell Edr Erdkunde* 1, 2, 1-26.

____1939 Das Pflanzenkleid des Nanga Parbat Wissen schaftliche Veroeffentlichungen des Deutschen Museums fuer Laenderkunde zu Leipzig (N.F) 7, 151-193 Vegetationskarte und 18,Bilden.

*Triestman J M 1972 *The Prehistory of China.* New Delhi.

Veevers J J Powell C Mc and Johnson B D 1975 Greater India's place in Gondwanaland and in Asia. Earth *Planet Sci* Lett 27,383- 387.

Virjee 1982 *Cytogeography of some Endemic Taxa of Kashmir Himalaya.* M Phil Dissert Univ of Kashmir Srinagar India.

____Dhar Uppeandra and Kachroo P 1985 Endemism IN *Frontier Areas of Plant Sciences (ed CP Malik et al) Narendra Publishing House Delhi 181-212.*

Walter H 1968 *Die Vegetation de erde* IN Oko-physiologischer Betrachtung 2. Die gemassigten and arktischen. Gustav Fischer Verlag Jena.

____1974 Die Vegetation Ostenrops and Zentralasiens. Stuttgart.

Wandelbo 1966 Trekk av Afghanistan plantgeografi. Acta Univ Gothob. *Botan Gothob* 5,1-20.

____1971 *Some distributional patterns within the flora Iranica area IN Plant Life of Southwest Asia.* 29-41 Edinburgh.

Weston W 1971 Cultural Frontiers of Ancient East Asia. Edinburgh

Yong Ma 1984 A silken highway from east to west. UNESCO Courier June 1984, 22-24.

Zalessky M D 1918 Mem Com-*Geol NS Livr* 174, 5-6.9

Zohary M 1963 On the geological structure of Iran. *Bull Res Counc Israel* 11 D Suppl 113.

____1971 The phytogeographical foundations of the Middle East IN Plant Life in southwest Asia. 42-52. Edinburgh

____1973 *Geobotanical Foundations of the Middle* East 1,2. Stuttgart-Amsterdam

---

* IN Current Trends in Geology (eds DP Agrawal Sheela Kusumgar and R V Krishnamurthy)-Delhi.

+ not seen in original.

INDEX

H

I

J

K

L

M